First published 1970
by Methuen Educational Limited
11 New Fetter Lane, London EC4

Filmset by Keyspools Ltd, Golborne, Lancs.
Printed in Great Britain by
Cox and Wyman Ltd, Fakenham, Norfolk

SBN 423 44160 4 net edition
423 79710 7 non-net edition

Distributed in the USA
by Barnes and Noble Inc

Contents

Preface

This book is written for students who study vectors at Sixth Form level. Much of the material has already been extensively tested in the classroom and pupils have found the content valuable both in their development of the subject and in their preparation for the various examining boards.

In this connection, I wish to thank the Oxford and Cambridge Schools Examinations Board (*O & C*), the University of Cambridge Local Examinations Syndicate (*C*, *CS*) and the University of London School Examinations Council (*L*) for their generous offers to quote questions from recent examinations.

Students will find the book particularly useful in

1 the numerous examples and problems set throughout
2 both the geometric and dynamic work given
3 the set of revision papers and
4 the many hints and complete set of answers printed at the end.

Finally, I must thank the many students at Forest Grammar School who have conscientiously worked through the problems, the artist who has so clearly illustrated the text and Methuen Educational who have taken such great care over the publishing.

D.L.A.H.

1 · Vectors

Directed line segments

Can we plot the course of a ship which steams at 10 km h^{-1} for 1 hour?

No: this information tells us only that the ship's position is somewhere on or within the circumference of a circle, radius 10 km.

Suppose, however, that the ship steams on a bearing 030°, that is North 30° East. We now know its course and its position is at A, say, where A is 10 km from O in the direction 030°. The displacement, that is the change of position, of the ship is shown in Figure 1.1.

Figure 1.1

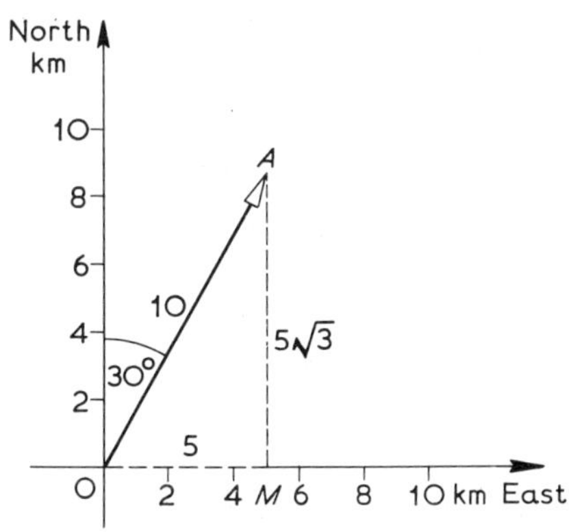

The ship has travelled from O to A, indicated by the arrow, and its displacement is represented by $\overrightarrow{OA}$, a *directed line segment*.

The length OM is the x component of $\overrightarrow{OA}$ and MA is the y component. Hence $\overrightarrow{OA}$ has components $(5, 5\sqrt{3})$.

If at A the ship alters course to steam due East at 10 km h^{-1} for 1 hour, then we can represent its new course by the directed line segment $\overrightarrow{AB}$ (Figure 1.2).

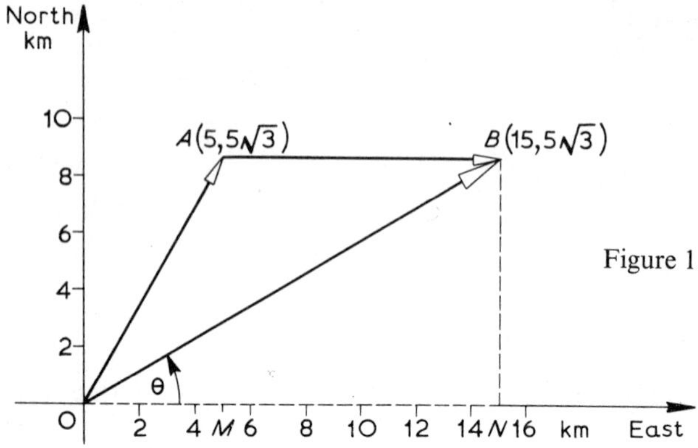

Figure 1.2

From the figure we see that the coordinates of B are

$$x = OM + MN = 5 + 10 = 15 \quad \text{and} \quad y = MA = NB = 5\sqrt{3}.$$

The result of the two displacements $\overrightarrow{OA}$ followed by $\overrightarrow{AB}$ is, therefore, $\overrightarrow{OB}$, with components $(15, 5\sqrt{3})$.

We write the statement, '$\overrightarrow{OA}$ followed by $\overrightarrow{AB}$ results in $\overrightarrow{OB}$', as

$$\overrightarrow{OA} + \overrightarrow{AB} = \overrightarrow{OB}$$

The + symbol here has a special meaning: add the respective x and y components of the directed line segments so that

$$\overrightarrow{OA} + \overrightarrow{AB} = \Big((5+10), (5\sqrt{3}+0)\Big) = (15, 5\sqrt{3}) = \overrightarrow{OB}$$

The distance of the ship from O is, by Pythagoras' theorem,

$$\sqrt{ON^2 + NB^2} = \sqrt{15^2 + (5\sqrt{3})^2} = \sqrt{225 + 75} = \sqrt{300}$$

$$= 10\sqrt{3} \text{ km}.$$

The direction $\overrightarrow{OB}$ makes with the East is given by $\tan\theta = \dfrac{NB}{ON} = \dfrac{5\sqrt{3}}{15}$ whence $\theta = 30°$, so that the ship is now on a bearing 060° from O.

Mathematicians, physicists and navigators have long been accustomed to represent the motion of objects by directed line segments because displacements, velocities, accelerations and forces are quantities which combine both magnitude and direction in their definitions.

Exercise 1

In the following questions $A = (3, 4)$, $B = (5, 12)$, $C = (6, 6)$ and $D = (8, 15)$.

1 Find the lengths and directions of the line segments (i) $\overrightarrow{OA}$, $\overrightarrow{OB}$ and $\overrightarrow{AB}$, (ii) $\overrightarrow{OC}$, $\overrightarrow{OD}$ and $\overrightarrow{CD}$, (iii) $\overrightarrow{AC}$, $\overrightarrow{BD}$ and $\overrightarrow{AD}$.

2 Write down the coordinates of A_1 and B_1 where $A_1 = 3A$ and $B_1 = \frac{1}{5}B$.
What are the lengths and directions of $\overrightarrow{OA_1}$ and $\overrightarrow{OB_1}$?

3 Write down the coordinates of C_1 and D_1 where $C_1 = \frac{1}{2}C$ and $D_1 = 2D$.
What are the lengths and directions of $\overrightarrow{OC_1}$ and $\overrightarrow{OD_1}$?

4 What are the coordinates of the mid-points of (i) $\overrightarrow{OA}$, $\overrightarrow{OB}$ and $\overrightarrow{AB}$, (ii) $\overrightarrow{OC}$, $\overrightarrow{OD}$ and $\overrightarrow{CD}$?

5 What are the components of the following resultants?
(i) $\overrightarrow{OA}+\overrightarrow{OB}$, (ii) $\overrightarrow{OC}+\overrightarrow{OD}$, (iii) $\overrightarrow{OA}+\overrightarrow{OB}+\overrightarrow{OC}$, (iv) $\overrightarrow{OA}+\overrightarrow{OB}+\overrightarrow{OC}+\overrightarrow{OD}$.

6 (i) If $\overrightarrow{OA}+\overrightarrow{OE} = (9, 12)$, find $\overrightarrow{OE}$. What is $\overrightarrow{OE}$ in terms of $\overrightarrow{OA}$?
(ii) If $\overrightarrow{OB}+\overrightarrow{OF} = (7\frac{1}{2}, 18)$, find $\overrightarrow{OF}$. What is $\overrightarrow{OF}$ in terms of $\overrightarrow{OB}$?

7 Draw the directed line segments $\overrightarrow{OA}$ and $\overrightarrow{OB}$. Draw the resultant directed line segment $\overrightarrow{OR} = \overrightarrow{OA}+\overrightarrow{OB}$. Complete the figure $OARB$ and verify that the magnitude and direction of $\overrightarrow{OA} = \overrightarrow{BR}$ and similarly for $\overrightarrow{OB}$ and $\overrightarrow{AR}$.

8 Repeat question 7 for the directed line segments $\overrightarrow{OC}$ and $\overrightarrow{OD}$.

9 If the number pair (a, o) represents a kg of apples and o oranges, what is the resultant number pair of the following sum?

$$(1, 6)+(0, 4)+(4\tfrac{1}{2}, 0)+(3, 10)$$

Interpret your result.

10 The number triple (b, r, s) represents b basket balls, r rugby balls and s soccer balls. Five schools order the following quantities:

$A(4, 12, 0)$ $B(0, 8, 8)$ $C(3, 10, 2)$ $D(7, 0, 11)$
$E(5, 10, 12)$

What is the total quantity ordered?
Which school probably does not play (i) basket ball, (ii) rugby, (iii) soccer?

Directed line segments and Vectors

Because scientists found it convenient to represent those quantities which embody both magnitude and direction in their composition, a new algebra was invented. It is called *vector algebra* and is a direct consequence of the behaviour of certain physical phenomena observed in the natural world. The entities which display these double characteristics of magnitude and direction are called *vectors,* from the Latin verb *vehere* to carry, since they are so closely connected with the idea of movement.

We must state at once that to a mathematician such a description of vectors is too limited because he will also use vectors when direction is not defined. A mathematician calls anything a vector that satisfies the set of axioms on page 40. We shall develop our study of vectors, however, from the properties of directed line segments.

A vector, therefore, is a number pair $\mathbf{r} = [x, y]$, with its own special algebra. We shall always reserve square brackets for vectors to distinguish them from directed line segment components or point coordinates.

The components of a vector are the respective x and y lengths and these components define both the magnitude and direction of the vector.

The magnitude is $|\mathbf{r}| = \sqrt{x^2+y^2}$ and the direction is $\theta = \tan^{-1}(y/x)$.

The modulus of $\mathbf{r}$, that is $|\mathbf{r}|$, the magnitude of $\mathbf{r}$, we shall write as r.

The angle θ is between the positive direction of the x axis and the vector $\mathbf{r}$.

This definition is satisfactory for what is known as a tied or bound vector where the initial point of the vector is at the origin, O. We shall also find it necessary to employ free vectors where the initial point of the vector is not specified.

Free vectors

In Figure 1.3, **AB** is a vector which begins at $A(6, 1)$ and ends at $B(8, 5)$.

To describe **AB**, we give it the components which characterise the directed line segment $\overrightarrow{AB}$: $\mathbf{AB} = [8-6, 5-1] = [2, 4]$.

What does such a definition imply?

It means that we can have many directed line segments situated in the plane yet all associated with one vector $\mathbf{r}$ (Figure 1.4).

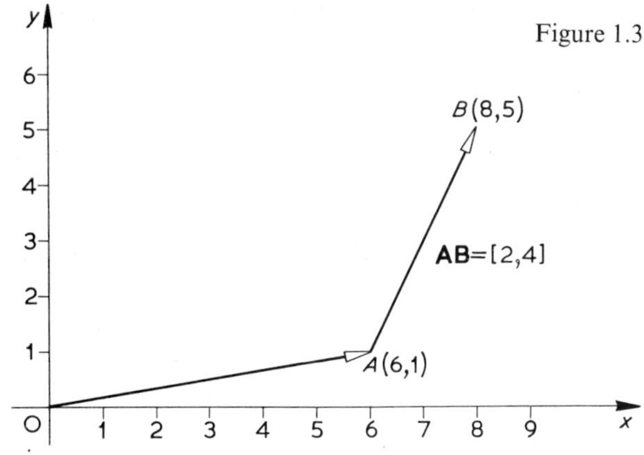

Figure 1.3

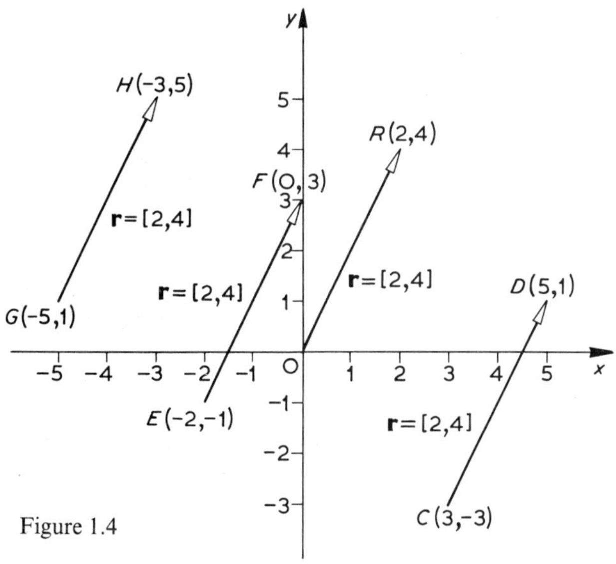

Figure 1.4

Though a vector describes the magnitude and direction of a directed line segment, it does not position it uniquely in the plane. Vectors, therefore, are not necessarily tied; they have no fixed position.

It is important at this stage to emphasize the point that vectors can be freely placed. Though the directed line segments of Figure 1.4 are all differently situated on the grid, they are, nevertheless, represented by the same vector. Should we choose to represent the directed line segment $\overrightarrow{GH}$ by the vector **s**, then

$$\mathbf{s} = [2, 4] \qquad \text{and} \qquad \mathbf{r} = [2, 4].$$

There is nothing in the respective components of **r** and **s** to distinguish one vector from the other. Both have the same magnitude

$$\sqrt{2^2+4^2} = \sqrt{20} = 2\sqrt{5} \text{ units}$$

and both have the same direction, given by $\tan\theta = 4/2 = 2$.

Therefore $\mathbf{r} = \mathbf{s}$: the vectors are *equivalent*; that is they both have the same magnitude and the same direction. In such cases we can interchange the symbols describing the vectors.

Figure 1.5

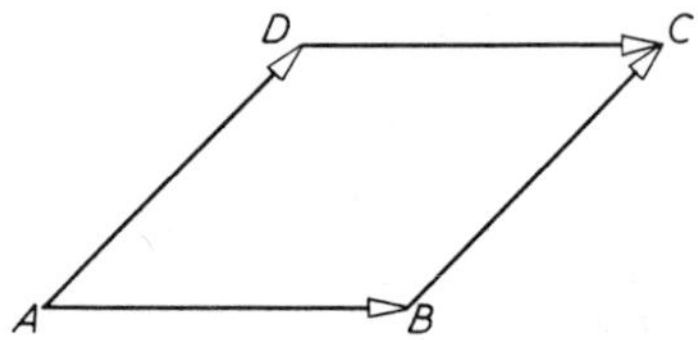

For example, in the rhombus of Figure 1.5, vector **AD** is of magnitude 2 units towards the North-east and **BC** is also of magnitude 2 units towards the North-east. Therefore **AD** = **BC** and we can write **AD** for **BC** or vice-versa.

Further, **AB** is of magnitude 2 units towards the East and **DC** is also of magnitude 2 units towards the East. Therefore **AB** = **DC**.

However, although the magnitudes of **AB** and **BC** are equal, **AB** $\neq$ **BC** because their directions are different.

Exercise 2

In this exercise, regard the vectors as directed line segments.

This diagram shows three vectors **a**, **b** and **c** drawn on a grid of unit squares.

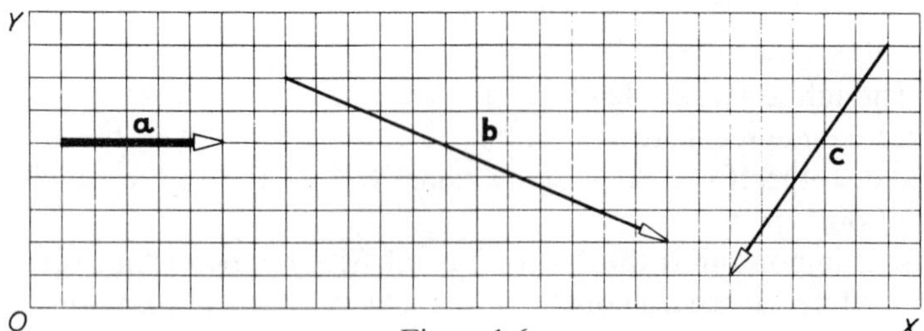

Figure 1.6

1 What are the magnitudes of the vectors **a**, **b** and **c**?
2 What angles do the vectors make with the lines parallel to OX?
3 Draw the following resultants on a similar grid:
(i) $\mathbf{a}+\mathbf{b}$ (ii) $\mathbf{a}+\mathbf{c}$ (iii) $\mathbf{b}+\mathbf{c}$
4 What are the magnitudes of the resultants in question 3?
5 What are the directions of the resultants in question 3?
6 The points $O(0, 0)$, $P(3, 4)$, $Q(-1, 2)$, $R(-2, -3)$ and $S(1, -1)$ are in the Cartesian plane. Let $\mathbf{p} = \mathbf{OP}$, $\mathbf{q} = \mathbf{OQ}$, $\mathbf{r} = \mathbf{OR}$ and $\mathbf{s} = \mathbf{OD}$. Construct the following vector resultants. In every question begin at the origin O, give the end point of each resultant and check your answer by adding the respective components. (i) $\mathbf{p}+\mathbf{q}$, (ii) $\mathbf{q}+\mathbf{p}$, (iii) $\mathbf{p}+\mathbf{s}$, (iv) $\mathbf{r}+\mathbf{s}$, (v) $\mathbf{s}+\mathbf{r}$, (vi) $\mathbf{s}+\mathbf{p}$, (vii) $\mathbf{q}+\mathbf{r}$, (viii) $\mathbf{s}+\mathbf{q}$, (ix) $\mathbf{s}+\mathbf{p}$.
7 What are the magnitudes of the vectors **p**, **q**, **r** and **s** in question 6?
8 What are the magnitudes of the resultants in question 6?
9 What are the directions of the vectors **p**, **q**, **r** and **s**?
10 What are the directions of the vector resultants in question 6?
11 Are there any equivalent vector resultants in question 6? If so, which are they?

Vector addition

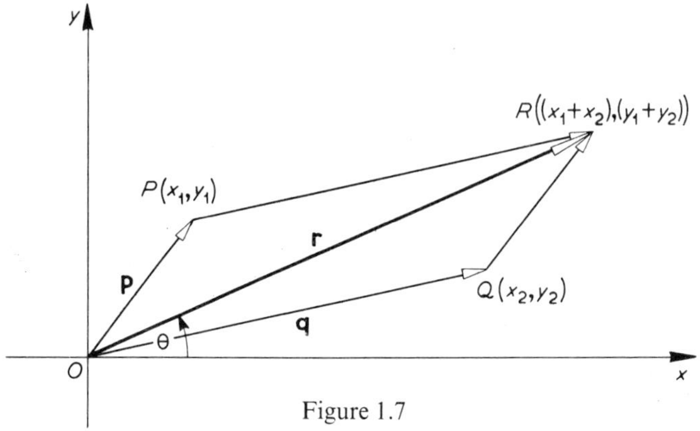

Figure 1.7

Consider the vectors **p** and **q** in Figure 1.7 where the coordinates of P and Q are (x_1, y_1) and (x_2, y_2) respectively.
When we add the respective components of **p** and **q**, then

$$\mathbf{p}+\mathbf{q} = [(x_1+x_2), (y_1+y_2)]$$

Let these be the components of vector **r**, so that $\mathbf{p}+\mathbf{q} = \mathbf{r}$.

From the figure we see that the magnitude of **r** is

$$\sqrt{(x_1+x_2)^2+(y_1+y_2)^2}$$

and the direction of **r** is given by

$$\tan\theta = \frac{y_1+y_2}{x_1+x_2}$$

What are the magnitude and direction of **QR**?

The length of **QR** is

$$\sqrt{(x_1+x_2-x_2)^2+(y_1+y_2-y_2)^2} = \sqrt{x_1^2+y_1^2}$$

The direction of **QR** is

$$\tan^{-1}\left(\frac{y_1+y_2-y_2}{x_1+x_2-x_2}\right) = \tan^{-1}\left(\frac{y_1}{x_1}\right)$$

The vector **QR**, therefore, is the same length and is in the same direction as **p**. Hence **p** is equivalent to **QR**.

Similarly **PR** is equivalent to **q**.

The figure *OPRQ* is, therefore, a parallelogram whose diagonal is **r**.

It is because all such additions of pairs of vectors result in the formation of parallelograms that this operation has become embodied in the *parallelogram law*:

When two adjacent sides of a parallelogram represent vectors, then the sum of these vectors is represented by the diagonal of the parallelogram drawn from the intersection of the two given sides.

This is a fundamental result.

From the figure also we can see that vector addition is commutative because whether we have $\mathbf{OP}+\mathbf{PR}$ or $\mathbf{OQ}+\mathbf{QR}$, that is $\mathbf{p}+\mathbf{q}$ or $\mathbf{q}+\mathbf{p}$, the result is always **OR**, that is **r**.

It is useful to notice that the addition of vectors **p** and **q** results in the vector **r** where **r** is the third side of the triangle *OPQ* formed by the three vectors.

We must, however, be careful to remember that it is the sum of the vectors **p** and **q** that is equal to **r** and not the sum of the magnitudes of **p** and **q** that equals the magnitude of **r**. Except in one case, we have $|\mathbf{p}|+|\mathbf{q}| > |\mathbf{r}|$.

When does that exception occur? When the triangle degenerates into a straight line so that **p** and **q** are collinear and $|\mathbf{p}|+|\mathbf{q}| = |\mathbf{r}|$.

Sum of many vectors

Suppose a horse runs from A_0 to A_1, represented by $\mathbf{a}_1$ (Figure 1.8), from A_1 to A_2, $\mathbf{a}_2$, and then on to A_3, $\mathbf{a}_3$. What is its resultant displacement from A_0?

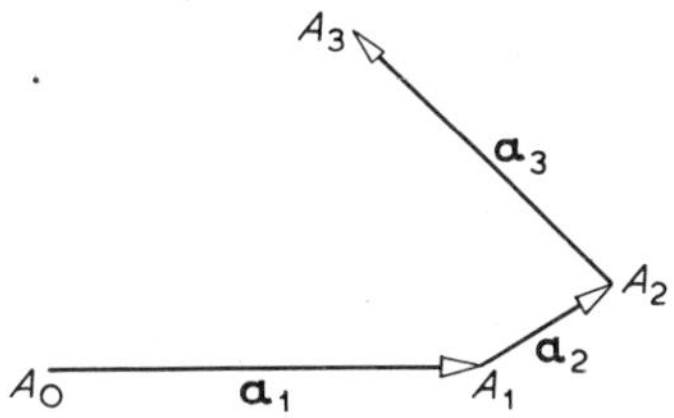

Figure 1.8

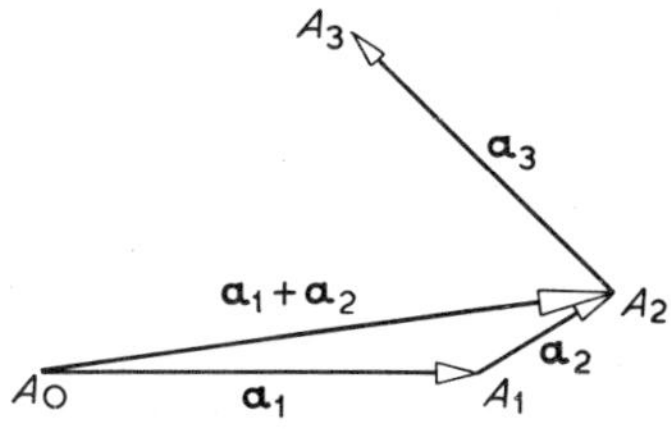

Figure 1.9

We can first say that $\mathbf{a}_1+\mathbf{a}_2 = \mathbf{A}_0\mathbf{A}_2$ (Figure 1.9). If we now take triangle $A_0A_2A_3$, we next obtain $(\mathbf{a}_1+\mathbf{a}_2)+\mathbf{a}_3 = \mathbf{A}_0\mathbf{A}_3$ (Figure 1.10).

The resultant displacement, therefore, is the vector sum $\mathbf{a}_1+\mathbf{a}_2+\mathbf{a}_3 = \mathbf{A}_0\mathbf{A}_3$.

Figure 1.10

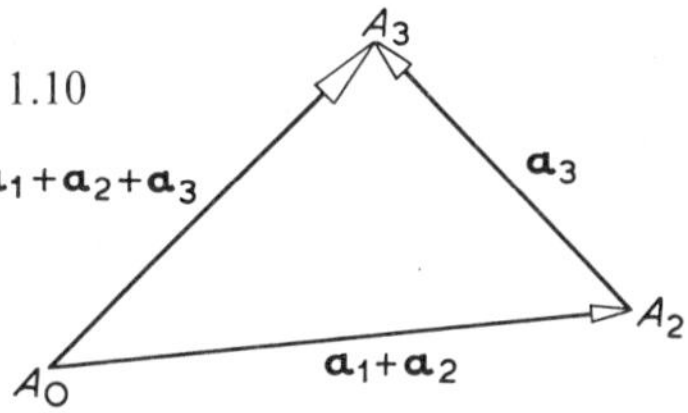

Suppose now the horse is in a point-to-point race from A_0 to A_1, on to A_2, to A_3, A_4, ... to A_n. If the corresponding displacements are $\mathbf{a}_1, \mathbf{a}_2, \mathbf{a}_3, \mathbf{a}_4, \ldots \mathbf{a}_n$, then by extending the above argument, we find that the resultant displacement is $\mathbf{a}_1+\mathbf{a}_2+\mathbf{a}_3+\mathbf{a}_4 \ldots +\mathbf{a}_n = \mathbf{A}_0\mathbf{A}_n$. This resultant vector completes a polygon of vectors. Should the polygon be closed by $\mathbf{a}_n$ then the resultant displacement is zero.

We are not, therefore, restricted to two vectors at a time but can have as many as we like. If they all chase one another, follow-my-leader style, then their resultant is simply the straight line join from the start, A_0, to the finish, A_n, that is $\mathbf{A}_0\mathbf{A}_n$.

Order of addition

Does it matter in what order we add vectors?

For example, in Figure 1.8, do we obtain the same resultant vector for the addition of $\mathbf{a}_3$ to $(\mathbf{a}_1+\mathbf{a}_2)$ as we do for the addition of $(\mathbf{a}_2+\mathbf{a}_3)$ to $\mathbf{a}_1$? That is, is it true that $(\mathbf{a}_1+\mathbf{a}_2)+\mathbf{a}_3 = \mathbf{a}_1+(\mathbf{a}_2+\mathbf{a}_3)$?

We know from Figures 1.9 and 1.10 that when we add together $\mathbf{a}_1$ and $\mathbf{a}_2$ first and then add $\mathbf{a}_3$, the resultant vector is $\mathbf{OA}_3$.

Let us now take the sum $(\mathbf{a}_2+\mathbf{a}_3)$ (Figure 1.11), that is $\mathbf{A}_1\mathbf{A}_3$, and add this to $\mathbf{a}_1$. We obtain Figure 1.12, where $\mathbf{a}_1+(\mathbf{a}_2+\mathbf{a}_3)$ is still $\mathbf{A}_0\mathbf{A}_3$, that is $\mathbf{a}_1+\mathbf{a}_2+\mathbf{a}_3$.

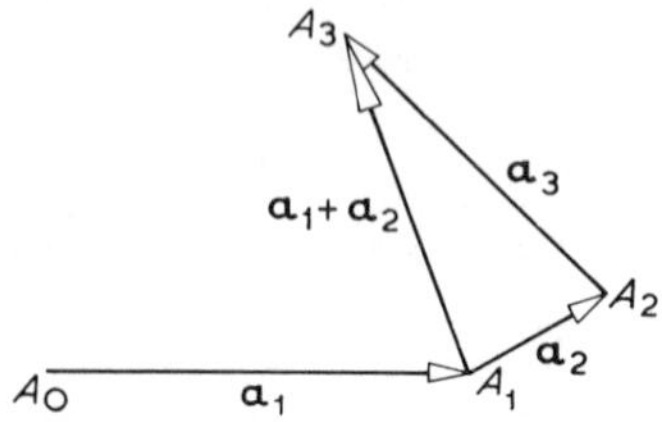

Figure 1.11

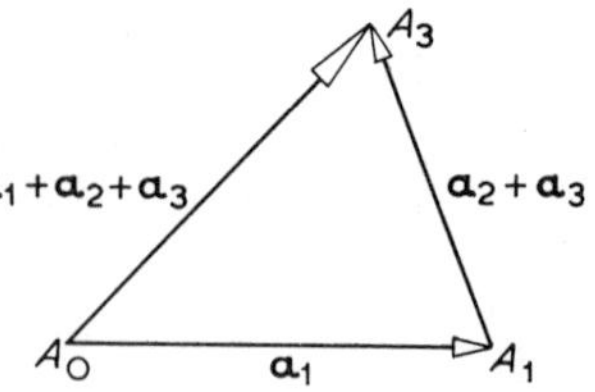

Figure 1.12

The order of addition is, therefore, immaterial and so vectors obey the associative law.

Example 1

What is the resultant of two forces $\mathbf{F}_1 = [5, 2]$ and $\mathbf{F}_2 = [1, 4]$ acting through the point O?

If the resultant force is $\mathbf{F}$, then $\mathbf{F} = [(5+1), (2+4)] = [6, 6]$.

Figure 1.13

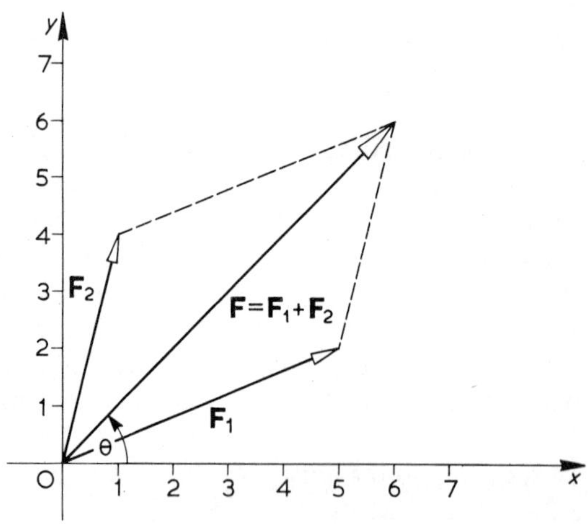

Figure 1.13 shows the two forces and their resultant. The figure indicated is the parallelogram of forces whose diagonal represents $\mathbf{F}$.

The magnitude of $\mathbf{F}$ is $\sqrt{6^2+6^2} = 6\sqrt{2}$ units and the direction of $\mathbf{F}$ is given by $\tan\theta = 6/6 = 1$, whence $\theta = 45°$.

Example 2

An aircraft can fly at 800 km h^{-1} in still air. What is its resultant velocity if it flies at this speed and is set on a course due West in a wind which blows from the North-east at 120 km h^{-1}? Further, what is the angle of drift?

The speed of an aircraft in still air is called its air-speed. This speed is relative to the air and because of air currents it is different from the aircraft's speed relative to the ground. The ground-speed and track of the aircraft are, therefore, also dependent upon the wind.

The velocity of the aircraft relative to the ground is the resultant of its velocity relative to the air and the velocity of the air, that is the wind, relative to the ground. The situation is represented by this equation

$$_{\text{aircraft}}\mathbf{v}_{\text{ground}} = {}_{\text{aircraft}}\mathbf{v}_{\text{wind}} + {}_{\text{wind}}\mathbf{v}_{\text{ground}} \quad \text{or } \mathbf{AG} = \mathbf{AW} + \mathbf{WG}$$

Notice the order of the letters: the inner letters, *W* and *G*, of the right-hand side of the equation are absorbed to leave only the extremes, *A* and *G*. We can show this by a vector triangle (Figure 1.14).

Figure 1.14

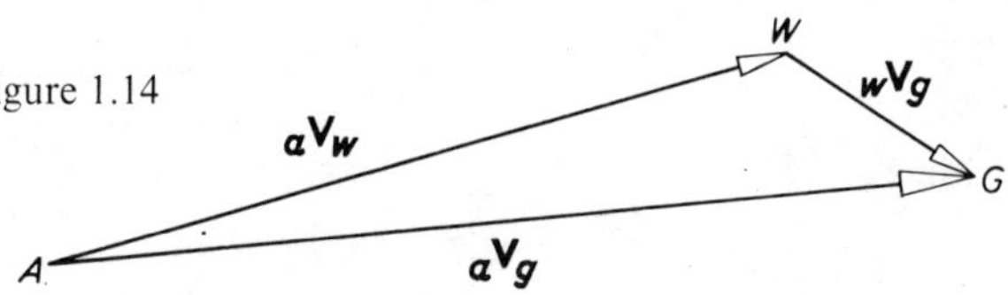

In this problem, we choose a suitable scale such as 10 mm to 100 km h^{-1} and represent the air-speed of 800 km h^{-1} in the direction due West by **AW** (Figure 1.15).

Figure 1.15

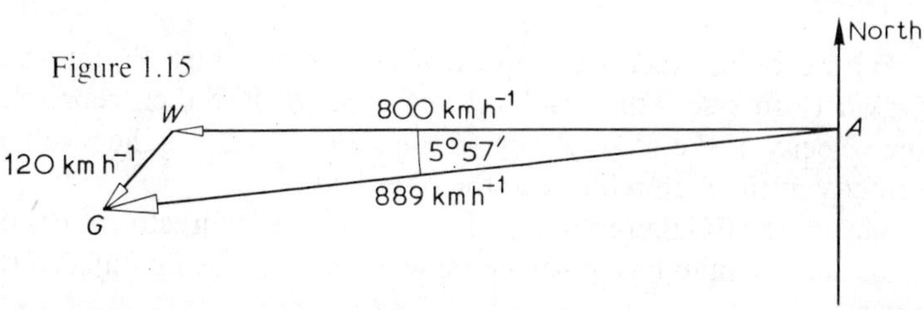

The velocity of the wind is represented by **WG** and the resultant track and ground-speed of the aircraft is then given by **AG**.

The resultant velocity, **AG**, is nearly 889 km h^{-1} in the direction South 84° 3′ West. The angle of drift, therefore, is 5° 57′.

Example 3

A river flows at 1 m s^{-1} and a man can row in still water at $1\frac{2}{3}$ m s^{-1}. He wants to row directly across the river. In what direction must he set course? How long does he take to cross the river if it is $133\frac{1}{3}$ m wide?

This problem is similar to example 2. Speed in still water means the speed of the boat relative to the water so that when there is a current flowing, the 'true' velocity of a boat is its velocity relative to the ground over which it is moving. This velocity is the resultant of the velocity of the boat relative to the water and the velocity of the water, the current, relative to the ground, that is

$${}_{\text{boat}}\mathbf{v}_{\text{ground}} = {}_{\text{boat}}\mathbf{v}_{\text{water}} + {}_{\text{water}}\mathbf{v}_{\text{ground}} \quad \text{or} \quad \mathbf{BG} = \mathbf{BW} + \mathbf{WG}.$$

With the scale 30 mm to 1 m s^{-1}, we draw the vector **WG** to represent the flow of the river (Figure 1.16). Since the man wishes to go straight across the stream, a line of convenient but indeterminate length is drawn perpendicularly to **WG**. A segment of this line represents the resultant.

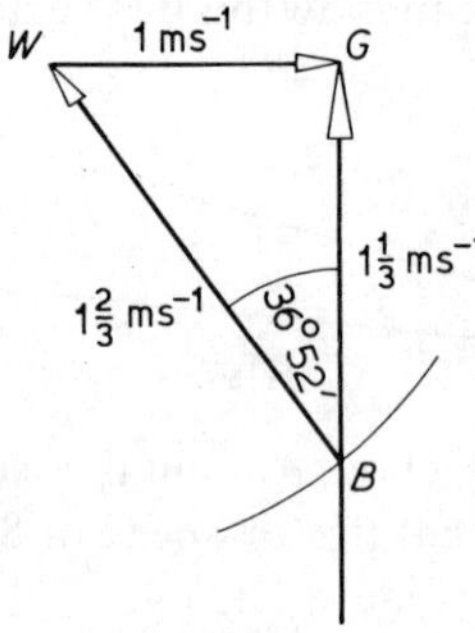

Figure 1.16

With a radius vector to represent $1\frac{2}{3}$ m s^{-1}, centre W, an arc is drawn to intersect the resultant vector at B. **BW** then represents the velocity of the boat relative to the water and **BG** is the resultant velocity of the man relative to the river bed.

The vector **BG** represents a velocity of $1\frac{1}{3}$ m s^{-1} directly across the river and the man has to set a course 53° 8′ with the up-bank of the river.

Since the river is $133\frac{1}{3}$ m wide, the time taken to cross is $133\frac{1}{3}/1\frac{1}{3}$ seconds, that is 100 seconds or 1 minute 40 seconds.

Exercise 3

1 Let $\mathbf{a} = [3, 4]$, $\mathbf{b} = [-1, 2]$, $\mathbf{c} = [-2, -3]$ and $\mathbf{d} = [1, -1]$. Find the following vector resultants: (i) $\mathbf{c}+\mathbf{a}+\mathbf{b}$, (ii) $\mathbf{b}+\mathbf{c}+\mathbf{a}$, (iii) $\mathbf{d}+\mathbf{b}+\mathbf{a}$, (iv) $\mathbf{a}+\mathbf{b}+\mathbf{c}+\mathbf{d}$, (v) $\mathbf{d}+\mathbf{a}+\mathbf{c}+\mathbf{b}$.

2 What are the magnitudes and directions of the vector resultants in question 1?

3 What are the equivalent vector resultants in question 1?

In the next three questions, find the magnitude and direction of the vector sum.

4 A velocity of 3 m s^{-1} West and 4 m s^{-1} South.

5 An acceleration of 2 m s^{-2} East and 2 m s^{-2} North.

6 A force of 2 newtons East and 3 newtons South-east.

7 The pilot of an aircraft sets a course due East and flies at a steady height at a speed of 800 km h^{-1}. The wind velocity is 80 km h^{-1} from the North. Find the resultant velocity of the aircraft (*a*) graphically and (*b*) by calculation.

8 A man can row a boat in still water at 8 km h^{-1}. Find the velocity of the boat if the man heads directly across a river which flows at 6 km h^{-1}.

9 An aircraft can fly in still air at 1200 km h^{-1}. If the pilot sets a course 065°, and the wind velocity is 80 km h^{-1} from 320°, find graphically the resultant velocity of the aircraft.

10 A pilot wants to go due South in an aircraft which can fly at 1100 km h^{-1} in still air. If the wind vector is 100 km h^{-1} from the North-east, what course must the pilot set?

11 A boy can swim at 1 m s^{-1} in still water. If he swims directly across a river which is flowing at $1\frac{1}{3}$ m s^{-1} what is his resultant velocity relative to the river bed? If the river is 100 m wide, how long does it take him to cross?

How far down-stream does the current take him?

12 In the quadrilateral *ABCD* when $\mathbf{AB} = \mathbf{DC}$, show that the figure is a parallelogram.

Multiplication by a number

If $\mathbf{p}$ is a displacement vector which represents a distance of 1 km due West, what vector represents a distance of 2 km due West?

Since 2 km is twice 1 km and the directions are the same, you have probably correctly suggested that the answer is twice $\mathbf{p}$, or $2\mathbf{p}$.

Suppose the distance is 10 km due West or $\frac{1}{2}$ km due West. What are the respective vectors now? Of course, they are $10\mathbf{p}$ and $\frac{1}{2}\mathbf{p}$.

What, now, does the vector $-\mathbf{p}$ represent?

Because we associate the magnitude of a vector with its modulus, $|\mathbf{p}|$, the magnitude of $-\mathbf{p}$ equals the magnitude of $\mathbf{p}$. Clearly, however, $-\mathbf{p}$ is different from $\mathbf{p}$ which must mean that its direction is different. You may have rightly guessed that the direction of $-\mathbf{p}$ is exactly opposite to that of $\mathbf{p}$ (Figure 1.17).

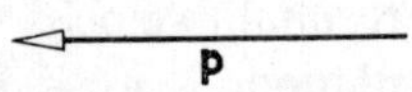

Figure 1.17

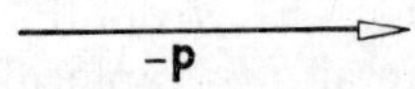

We define the vector $m\mathbf{p}$, therefore, to have magnitude $|m\mathbf{p}| = |m||\mathbf{p}|$, that is $|m|$ times that of p, to be in the direction of $\mathbf{p}$ when $m > 0$ and to be in the direction opposite to that defined by $\mathbf{p}$ when $m < 0$.

Example 1

If $\mathbf{p} = [2, 6]$ and $\mathbf{q} = [-2, 1]$ what is (i) $2\mathbf{p}+3\mathbf{q}$ and (ii) $2\mathbf{q}-\frac{1}{2}\mathbf{p}$?

(i) $2\mathbf{p}+3\mathbf{q} = 2[2, 6] \quad +3[-2, 1]$
$\quad = \quad [4, 12] \; + \; [-6, 3] = [-2, 15].$

(ii) $2\mathbf{q}-\frac{1}{2}\mathbf{p} = 2[-2, 1]-\frac{1}{2}[2, 6]$
$\quad = \quad [-4, 2]- \; [1, 3] \quad = [-5, -1].$

Example 2

Figure 1.18

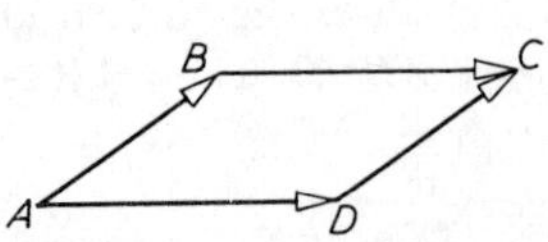

In a quadrilateral $ABCD$, if $\mathbf{AB} = \mathbf{DC}$, prove that $\mathbf{BC} = \mathbf{AD}$.
By the polygon of vectors, we have

$$\mathbf{AB}+\mathbf{BC}+\mathbf{CD} = \mathbf{AD} \qquad \ldots \qquad \ldots \text{(i)}$$

But $\quad \mathbf{AB} = \mathbf{DC}$

that is $\quad \mathbf{AB}-\mathbf{DC} = \mathbf{0} \quad$ or $\quad \mathbf{AB}+\mathbf{CD} = \mathbf{0}$

Hence (i) simplifies to **BC** = **AD**.

In this example, **AB** − **DC** = **0**. The symbol **0** is called the *zero vector* and so −**DC** or **CD** is the inverse of **AB**.

Before we consider more examples, we must determine whether

$$(\lambda+\mu)\mathbf{p} = \lambda\mathbf{p}+\mu\mathbf{p} \qquad \ldots \qquad \ldots (1)$$

and $\lambda(\mathbf{p}+\mathbf{q}) = \lambda\mathbf{p}+\lambda\mathbf{q} \qquad \ldots \qquad \ldots (2)$

where λ and μ are numbers.

For these two equations to be true we have to show that the left-hand side of each is equal in both magnitude and direction to the respective right-hand side.

In equation (1) there are many possibilities to consider, for example $\lambda > 0$, $\mu > 0$, $\mathbf{p} \neq \mathbf{0}$; $\lambda > 0$, $\mu < 0$, $\mathbf{p} \neq \mathbf{0}$ amongst them. We shall prove the first case and leave the other cases to be proved by the student.

On page 20 we have $|m\mathbf{p}| = |m||\mathbf{p}|$
so that $|\lambda\mathbf{p}|+|\mu\mathbf{p}| = |\lambda||\mathbf{p}|+|\mu||\mathbf{p}|$

Now λ and μ are both positive, so that we have

$$|\lambda\mathbf{p}|+|\mu\mathbf{p}| = (\lambda+\mu)|\mathbf{p}|$$

that is, the magnitudes of both sides of equation (1) are the same. The directions of $(\lambda+\mu)\mathbf{p}$ and both $\lambda\mathbf{p}$ and $\mu\mathbf{p}$ are the same because they are all in the direction of **p**.

Therefore $(\lambda+\mu)\mathbf{p} = \lambda\mathbf{p}+\mu\mathbf{p}$.

In equation (2) we consider the case when $\lambda > 0$ and leave the other possibilities to be considered by the student.

In triangle OPQ, let $\mathbf{OP} = \mathbf{p}$, $\mathbf{PQ} = \mathbf{q}$ and then $\mathbf{OQ} = \mathbf{p}+\mathbf{q}$ (Figure 1.19).

Figure 1.19

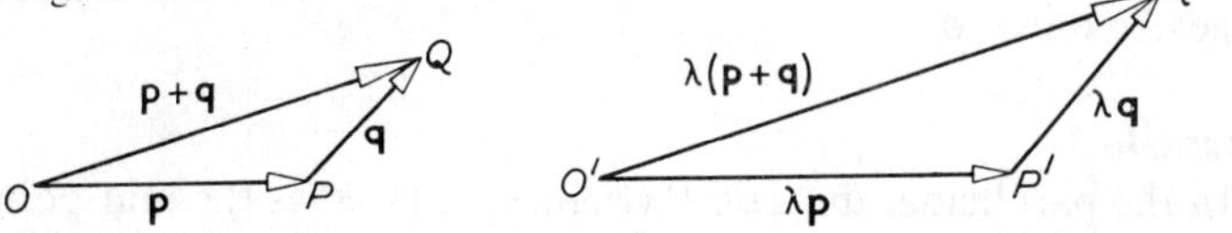

Triangle $O'P'Q'$ is similar to triangle OPQ and $\mathbf{O'P'} = \lambda\mathbf{p}$, so that $\mathbf{P'Q'} = \lambda\mathbf{q}$ and $\mathbf{O'Q'} = \lambda(\mathbf{p}+\mathbf{q})$. Therefore in triangle $O'P'Q'$, we have

$$\lambda\mathbf{p}+\lambda\mathbf{q} = \lambda(\mathbf{p}+\mathbf{q})$$

An important example

Figure 1.20

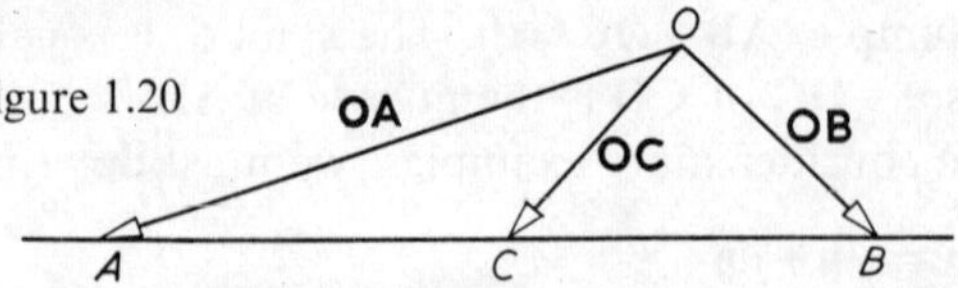

In Figure 1.20 prove that $\lambda\mathbf{OA}+\mu\mathbf{OB} = (\lambda+\mu)\mathbf{OC}$, where C is the point in AB such that $AC:CB = \mu:\lambda$.

The following proof, though not the quickest, illustrates many of the points we have discussed.

$$\mathbf{OA}+\mathbf{AC} = \mathbf{OC}$$

Therefore $$\lambda\mathbf{OA}+\lambda\mathbf{AC} = \lambda\mathbf{OC} \qquad \ldots \qquad \ldots \text{(i)}$$

Again $$\mathbf{OB}+\mathbf{BC} = \mathbf{OC}$$

Therefore $$\mu\mathbf{OB}+\mu\mathbf{BC} = \mu\mathbf{OC} \qquad \ldots \qquad \ldots \text{(ii)}$$

Add (i) and (ii)

$$\lambda\mathbf{OA}+\mu\mathbf{OB}+\lambda\mathbf{AC}+\mu\mathbf{BC} = (\lambda+\mu)\mathbf{OC} \qquad \ldots \text{(iii)}$$

But $$AC:CB = \mu:\lambda$$

and since these segments are in the same straight line and direction, we have

$$\lambda\mathbf{AC} = \mu\mathbf{CB} \qquad \text{or} \qquad \lambda\mathbf{AC}-\mu\mathbf{CB} = \mathbf{0}$$

or $$\lambda\mathbf{AC}+\mu\mathbf{BC} = \mathbf{0}$$

Using this result in (iii), we have

$$\lambda\mathbf{OA}+\mu\mathbf{OB} = (\lambda+\mu)\mathbf{OC} \qquad \ldots \qquad \ldots \text{I}$$

We often use this result and we give applications now and on pages 26 onwards.

Example 3

In the parallelogram $ABCD$ (Figure 1.21), M is the mid-point of AB and AC intersects DM at N. Prove that DM and AC trisect each other.

From triangle AMD

$$2(\tfrac{1}{2}\mathbf{p})+1\mathbf{q} = (2+1)\mathbf{AN} = 3\mathbf{AN} \qquad \ldots \qquad \ldots \text{(i)}$$

where $MN:ND = 1:2$, so that N trisects MD.

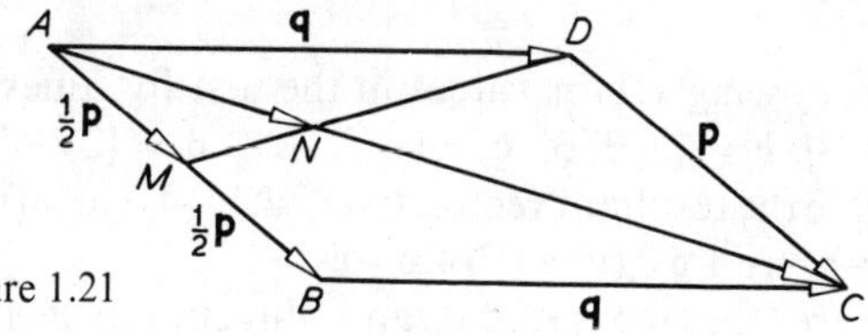

Figure 1.21

Again, from triangle ABC

$$\mathbf{p}+\mathbf{q} = \mathbf{AC}$$

Since, from (i), we know that

$$2(\tfrac{1}{2}\mathbf{p})+1\mathbf{q} = \mathbf{p}+\mathbf{q} = 3\mathbf{AN}$$

therefore $\mathbf{AC} = 3\mathbf{AN}$

This means that AC is in the same direction as AN and is 3 times as long. Therefore N trisects AC and so DM and AC trisect each other.

One further vector law will be required here. This law enables us to factorize or expand brackets as we do in ordinary algebra, for example

$$3(2\mathbf{p}+3\mathbf{q}) = 6\mathbf{p}+9\mathbf{q}$$

The law is given in the form

$$(\lambda\mu)\mathbf{p} = \lambda(\mu)\mathbf{p}$$

There are a number of possibilities. When none of the three quantities is zero, then the magnitudes of each side of the equation are equal to $|\lambda||\mu||\mathbf{p}|$.

The direction depends upon the signs of the multipliers. When the coefficients are either both positive or both negative, then the directions of both vectors on each side of the equation are the same as $\mathbf{p}$. When the coefficients have different signs, for example $\lambda > 0$ and $\mu < 0$, then both vectors have the direction defined by $-\mathbf{p}$.

Both sides of the equation, therefore, represent vectors which have the same magnitude and the same direction. Hence, $(\lambda\mu)\mathbf{p} = \lambda(\mu)\mathbf{p}$.

Exercise 4

Use the following vector values in the first five questions.

$\mathbf{a} = [4, 8]$, $\mathbf{b} = [-3, 6]$, $\mathbf{c} = [-5, -1]$, $\mathbf{d} = [2, -7]$

1 Calculate the resultant vector. (i) $\mathbf{b}-\mathbf{d}$, (ii) $\mathbf{d}-\mathbf{a}$, (iii) $\mathbf{a}+\mathbf{b}-\mathbf{c}-\mathbf{d}$, (iv) $\mathbf{a}+\mathbf{b}-(\mathbf{c}+\mathbf{d})$, (v) $\mathbf{a}-\mathbf{b}+\mathbf{c}-\mathbf{d}$

2 What are the magnitudes and directions of the resultant vectors in question 1?

3 Find the resultant vectors of the following:
(i) $3\mathbf{a}+4\mathbf{c}$, (ii) $2\mathbf{b}-5\mathbf{d}$, (iii) $3(\mathbf{a}+\mathbf{b}+\mathbf{c})$, (iv) $2\mathbf{a}-4\mathbf{b}+3\mathbf{c}$, (v) $\frac{1}{2}\mathbf{a}-\frac{2}{3}\mathbf{b}$, (vi) $\frac{1}{4}\mathbf{a}+\frac{1}{3}(\mathbf{c}+\mathbf{d})$, (vii) $\mathbf{b}+\dfrac{\mathbf{a}-\mathbf{c}}{3}$, (viii) $4(\mathbf{b}+\mathbf{c})-\mathbf{a}$.

4 Verify the following with the above component values:
(i) $3(2\mathbf{c}) = 6\mathbf{c} = 2(3\mathbf{c})$ (ii) $3(\mathbf{a}+\mathbf{b}) = 3\mathbf{a}+3\mathbf{b}$
(iii) $4(\mathbf{b}-\mathbf{d}) = 4\mathbf{b}-4\mathbf{d}$ (iv) $-2(\mathbf{a}+\mathbf{c}) = -2\mathbf{a}-2\mathbf{c}$
(v) $(5+2)\mathbf{d} = 5\mathbf{d}+2\mathbf{d}$ (vi) $(5-2)\mathbf{d} = 5\mathbf{d}-2\mathbf{d}$

5 Verify the following inequalities:
(i) $|\mathbf{a}+\mathbf{b}| \leqslant |\mathbf{a}|+|\mathbf{b}|$ (ii) $|\mathbf{a}-\mathbf{b}| \geqslant |\mathbf{a}|-|\mathbf{b}|$

6 A force of 10 newtons acts on an object at an angle of 60° to the x axis. What two forces acting in the positive x and y directions, respectively, have an equivalent effect on the object?

7 Three forces act on an object: $\mathbf{F}_1 = 2$ newtons at 030°, $\mathbf{F}_2 = 3$ newtons at 100° and $\mathbf{F}_3 = 6$ newtons at 250°. Find, graphically, their resultant.

8 In the triangle ABC, prove that $\mathbf{AB}+\mathbf{AC} = 2\mathbf{AM}$, where M is the mid-point of BC.

9 Prove that the line joining the mid-points of two sides of a triangle is equal in magnitude to one-half the third side and is parallel to it.

10 Prove that the figure formed by the join of the mid-points of the sides of a quadrilateral taken in order is a parallelogram.

11 Prove that the two lines from a vertex of a parallelogram to the mid-points of the opposite sides trisect the diagonal they cross.

12 Prove that the lines joining the mid-points of opposite sides of a quadrilateral bisect each other.

13 If O is any point in the plane of a triangle ABC, prove that

$$\mathbf{OA}+\mathbf{CB} = \mathbf{OB}+\mathbf{CA}$$

14 In the triangle ABC, the points P, Q and R divide the sides BC, CA and AB in the ratios 1:3, 3:1 and 1:1, respectively. Prove that

$$2\mathbf{AP}+2\mathbf{BQ}+\mathbf{CR} = \mathbf{0}$$

The vector b — a

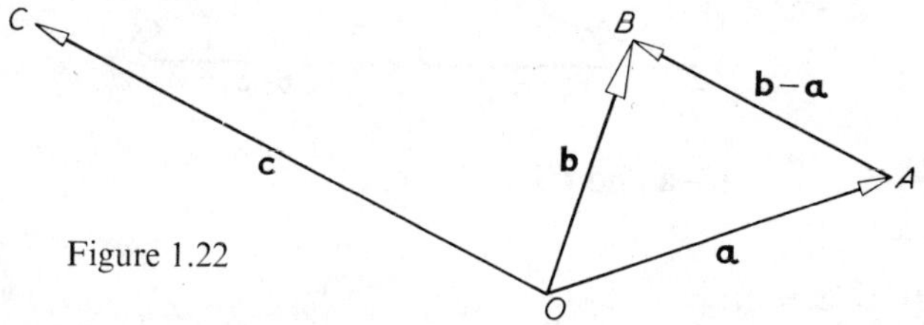

Figure 1.22

From a fixed point O we have already said that the point A may be reached by the displacement vector **a**. This is a convenient notation. In the diagram (Figure 1.22), points A, B and C are shown: let the vectors **OA**, **OB** and **OC** be represented by **a**, **b** and **c**, respectively.

With these small letters, we give a useful notation for the vector **AB**. By reversing the direction of vector **a**, we see that **AB** is the resultant of **AO**+**OB**, that is $-\mathbf{a}+\mathbf{b}$ or $\mathbf{b}-\mathbf{a}$. Therefore

$$\mathbf{AB} = \mathbf{b}-\mathbf{a}$$

Notice the order of the letters. Similarly $\mathbf{BC} = \mathbf{c}-\mathbf{b}$.

Parallel vectors

Suppose that **c** is twice the magnitude of **AB** and parallel to **AB**. We know from page 20 that $\mathbf{c} = 2(\mathbf{b}-\mathbf{a})$.

If we represent $\mathbf{b}-\mathbf{a}$ by **d**, then $\mathbf{c} = 2\mathbf{d}$ is an expression for parallel vectors.

We therefore have parallel vectors when

$$\mathbf{p} = m\mathbf{q}$$

for a number $m \neq 0$.

From this equation, we cannot say that $\mathbf{p}/\mathbf{q} = m$ because we have no definition for the division of vectors. There is in vector algebra no inverse for the operation of multiplication, although, as we noticed on page 21, subtraction is the inverse of addition.

Example

We now show a neater way to prove the important example given on page 22. Let $\mathbf{OA} = \mathbf{a}$, $\mathbf{OB} = \mathbf{b}$ and $\mathbf{OC} = \mathbf{c}$ (Figure 1.23).

Figure 1.23

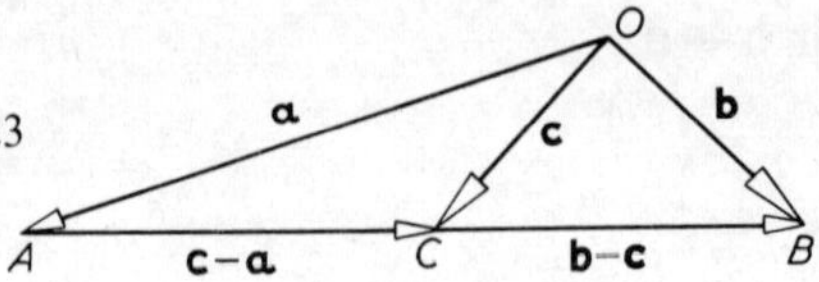

Therefore $\mathbf{AC} = \mathbf{c}-\mathbf{a}$ and $\mathbf{CB} = \mathbf{b}-\mathbf{c}$.

Since

$$AC:CB = \mu:\lambda$$

then

$$\lambda\mathbf{AC} = \mu\mathbf{CB}$$

that is

$$\lambda(\mathbf{c}-\mathbf{a}) = \mu(\mathbf{b}-\mathbf{c})$$

$$\lambda\mathbf{c}-\lambda\mathbf{a} = \mu\mathbf{b}-\mu\mathbf{c}$$

whence

$$(\lambda+\mu)\mathbf{c} = \lambda\mathbf{a}+\mu\mathbf{b} \qquad \ldots \qquad \ldots \text{I}$$

Why is the result of this example so important? For three immediate reasons:

(i) *Collinearity of Points*

We can write equation I as

$$\lambda\mathbf{a}+\mu\mathbf{b}-(\lambda+\mu)\mathbf{c} = \mathbf{0}$$

Now we know that A, C and B are collinear and that the coefficients λ and μ are numbers. What is the sum of these numbers?

$$\lambda+\mu-(\lambda+\mu) = \lambda+\mu-\lambda-\mu = 0$$

The sum is zero and we have, therefore, a condition for the points A, C and B to lie on a straight line: the sum of the vector coefficients must be zero.

The equation may be written in the form

$$l\mathbf{a}+m\mathbf{b}+n\mathbf{c} = \mathbf{0}$$

where $l = \lambda$, $m = \mu$ and $n = \lambda+\mu$.

Therefore if

$$l+m+n = 0$$

then the points A, B and C lie on a straight line.

The example, therefore, gives a condition for the collinearity of points.

(ii) *Line Segment Ratios*

Suppose we write equation I as

$$\mathbf{c} = \frac{\mu\mathbf{b}+\lambda\mathbf{a}}{\lambda+\mu}$$

Here $\mathbf{c}$ divides the line AB in the ratio $\mu:\lambda$. Let $\lambda = \mu = 1$. What is the significance of this?

$$\mathbf{c} = \frac{\mathbf{a}+\mathbf{b}}{2}$$

and this is the vector to position the mid-point of AB.

When $\mu:\lambda$ is negative the point C lies on either AB produced or BA produced according to which number is negative. Such a value gives an external division of AB.

Therefore the example gives a method to divide a line segment in a given ratio.

(iii) *Centroids*

The centroid, or mean centre, of two given points, whose positions are described by vectors $\mathbf{a}$ and $\mathbf{b}$, is determined by their mid-point vector, that is

$$\frac{\mathbf{a}+\mathbf{b}}{2}$$

The centroid is situated at the point G, given by $\mathbf{OG}$ or $\mathbf{g} = \dfrac{\mathbf{a}+\mathbf{b}}{2}$.

When there are n given points, then

$$\mathbf{g} = \frac{\mathbf{a}_1+\mathbf{a}_2+\mathbf{a}_3+\ldots+\mathbf{a}_n}{n} \qquad \ldots \qquad \ldots \text{II}$$

Again, when masses m_1, m_2, m_3, ... m_n are in the plane with corresponding position vectors $\mathbf{a}_1$, $\mathbf{a}_2$, $\mathbf{a}_3$, ... $\mathbf{a}_n$, we have the mass centre of the system at a point given by the vector

$$\mathbf{g} = \frac{m_1\mathbf{a}_1+m_2\mathbf{a}_2+m_3\mathbf{a}_3+\ldots+m_n\mathbf{a}_n}{m_1+m_2+m_3+\ldots+m_n} \qquad \ldots \qquad \ldots \text{III}$$

Centroid of the vertices of a triangle

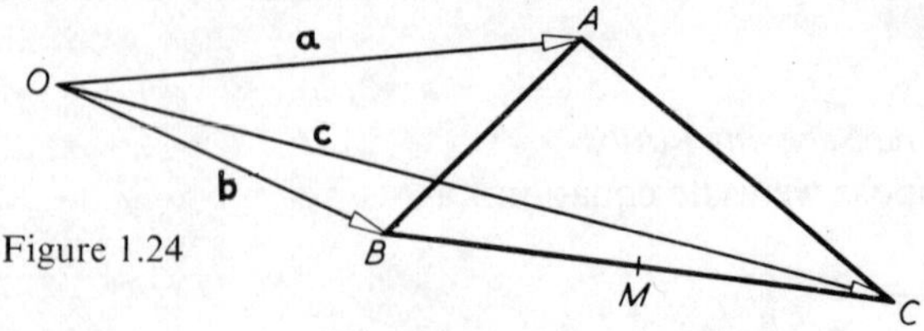

Figure 1.24

Let the vectors of the vertices of triangle ABC be **a**, **b** and **c** relative to a fixed point O (Figure 1.24).

From II the centroid of the vertices is at G whose vector is

$$\mathbf{g} = \frac{\mathbf{a}+\mathbf{b}+\mathbf{c}}{3}.$$

Where is this point G relative to A, B and C?

If we consider B and C only, then their mean centre is given by $\frac{\mathbf{b}+\mathbf{c}}{2}$, which is the mid-point M of BC.

The centroid of the triangle must now lie on AM, one of the medians, at the point G whose vector is given by **g**.

Let G divide AM in the ratio $\mu:\lambda$, that is $AG:GM = \mu:\lambda$. From I

$$\lambda\mathbf{a}+\mu\frac{\mathbf{b}+\mathbf{c}}{2} = (\lambda+\mu)\frac{\mathbf{a}+\mathbf{b}+\mathbf{c}}{3}$$

By comparing the coefficients of both sides of this equation, we see that $\lambda = 1$ and $\mu = 2$. Therefore $AG:GM = 2:1$.

Hence G trisects AM nearer to M.

Since $\mathbf{g} = \frac{\mathbf{a}+\mathbf{b}+\mathbf{c}}{3}$ is symmetrical in **a**, **b** and **c**, that is the expression is unaltered if we change the order of the letters, we have also proved that the centroid, G, lies on BN such that $BG:GN = 2:1$ and on CP where $CG:GP = 2:1$ (Figure 1.25).

That is the medians are concurrent at G, the common point of trisection.

Figure 1.25

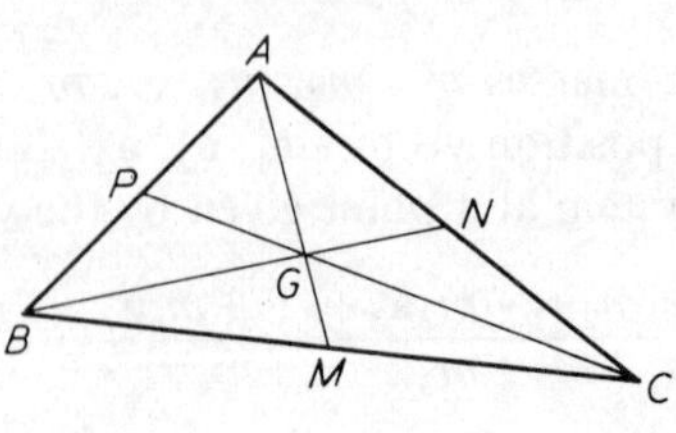

Example

A mass of 4 kg has its position described by vector [5, −3] and another mass of 2 kg has position vector [−4, 3]. Find the vector of the centre of mass.

If the vectors [5, −3] and [−4, 3] are **a** and **b**, respectively, then by III the vector of the centre of mass is

$$\mathbf{g} = \frac{4\mathbf{a}+3\mathbf{b}}{4+2} = \frac{4[5, -3]+2[-4, 3]}{6}$$

$$= \tfrac{1}{6}[12, -6] = [2, -1].$$

Relative velocity

If the velocity of A relative to a fixed point O is ${}_A\mathbf{v}_O$ and that of B is ${}_B\mathbf{v}_O$, then from page 17, we have

$${}_A\mathbf{v}_O = {}_A\mathbf{v}_B + {}_B\mathbf{v}_O$$

where ${}_A\mathbf{v}_B$ is the velocity of A relative to B.

If we know, therefore, the velocities of A and B relative to O, then we can find the velocity of A relative to B by the equation

$${}_A\mathbf{v}_B = {}_A\mathbf{v}_O - {}_B\mathbf{v}_O$$

That is ${}_A\mathbf{v}_B$ is the velocity of A together with the inverse velocity of B.

Similarly the velocity of B relative to A is ${}_B\mathbf{v}_A = {}_B\mathbf{v}_O - {}_A\mathbf{v}_O$.

By a similar argument, we can find the relative acceleration of, say, A to B

$${}_A\mathbf{a}_B = {}_A\mathbf{a}_O - {}_B\mathbf{a}_O$$

Example

The velocity of one ship A is 24 km h^{-1} due East and another ship B has velocity 18 km h^{-1} due North. What is the velocity of A relative to B?

$${}_A\mathbf{v}_B = {}_A\mathbf{v}_O - {}_B\mathbf{v}_O$$

$$= 24 \text{ km h}^{-1} \text{ due East} - 18 \text{ km h}^{-1} \text{ due North}$$

$$= 24 \text{ km h}^{-1} \text{ due East} + 18 \text{ km h}^{-1} \text{ due South}$$

Figure 1.26 shows the vector triangle.

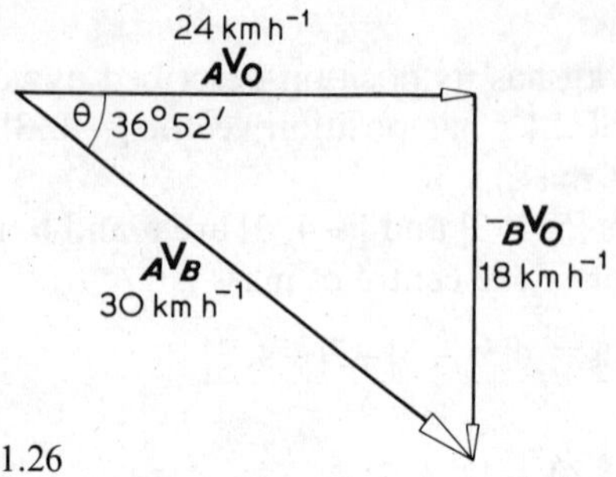

Figure 1.26

Because the triangle is right-angled, by Pythagoras' theorem the magnitude of the velocity of A relative to B is

$$|_A\mathbf{v}_B| = \sqrt{24^2+18^2} = 30 \text{ km h}^{-1}$$

The direction is given by $\tan\theta = 18/24 = 0{\cdot}75$, whence $\theta = 36°\ 52'$.

The resultant vector is the course and speed A appears to have to an observer on B.

Exercise 5

1 What is the vector of the centroid of the vertices of triangle ABC where $A = (-1, 4)$, $B = (8, -7)$ and $C = (5, 0)$?

2 What is the vector of the centre of mass of masses 3 kg and 2 kg at the points $(3, -4)$ and $(-2, 1)$, respectively?

3 The vertices of triangle ABC are $A = (0, 4)$, $B = (1, 1)$ and $C = (3, 3)$. If M is the mid-point of BC, find vectors **OM** and **AM**.

4 If A, P and B are three collinear points with vectors **a**, **p** and **b**, show that when $AP:PB = m:n$, then $\mathbf{p} = \dfrac{n\mathbf{a}+m\mathbf{b}}{m+n}$.

5 Show that when $3\mathbf{OA}+4\mathbf{OB}-7\mathbf{OC} = \mathbf{0}$, the points A, B and C are collinear.

6 If, in a quadrilateral $ABCD$, O is the mid-point of BD, prove that $\mathbf{AB}+\mathbf{AD}+\mathbf{CB}+\mathbf{CD} = 2(\mathbf{AO}+\mathbf{CO})$.
What condition put upon AO and CO ensures that the quadrilateral is a parallelogram?

7 Show that in a regular pentagon $ABCDE$, $\mathbf{AB}+\mathbf{BC}+\mathbf{AC}+\mathbf{AE}+\mathbf{ED}+\mathbf{AD} = 4\mathbf{AM}$ where M is the mid-point of CD.

8 Show that in a triangle ABC, $\mathbf{AB}+\mathbf{AC}+\mathbf{BC} = 4\mathbf{MN}$ where

M and N are the mid-points of AB and BC, respectively.

9 Find the resultant of $\mathbf{AB}+\mathbf{AC}+k\mathbf{BC}$ for a point M on AB such that $AM:MB = k:1$ and where N bisects BC in triangle ABC.

10 An aircraft flies on a course 030° for 100 km from A to B. It then flies to C a distance of 80 km at 060°. If the aircraft flies at a steady height find its displacement vector from A.

11 A ship sails from O to A a distance of 5 km in a direction 045°. It then alters course to sail to B a distance of 3 km at 160°. Find its displacement vector from O.

12 One ship A steams due South at a speed of 12 km h^{-1} and from it a ship B appears to be moving in a direction 315° at 5 km h^{-1}. Find the velocity of B.

13 An aircraft is flying due South at a speed of 1200 km h^{-1}. What is its velocity relative to another aircraft flying at the same height on a track 120° and at a speed of 700 km h^{-1}?

14 To a man walking due North at a speed of 6 km h^{-1}, the wind appears to come from the West. When he walks at 10 km h^{-1}, however, it appears to come from the North-west. Find the velocity of the wind.

15 Prove that the centroid of a system of particles, S, is independent of any fixed point of reference.

16 Prove that the centroid of two discrete systems of particles, S and S', is the centroid of the composite system of particles S and S'.

Vectors in three dimensions

Where is the centroid of a crystal?

What is the resultant acceleration of a mass moving under different forces in space?

How does energy vary in a magnetic field?

These questions imply three dimensions: how can we represent vectors in this space?

Indeed, need we be restricted to three dimensional space?

We have found it convenient so far to use the Cartesian system of coordinates (x, y). For three dimensions the system has coordinates (x, y, z) (Figure 1.27), where the axes are mutually perpendicular to one another.

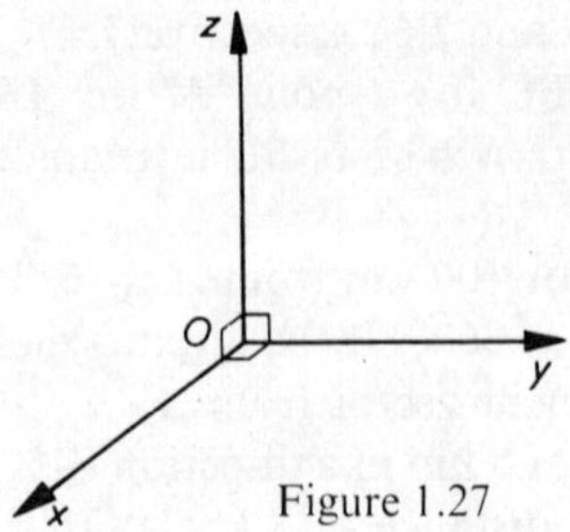

Figure 1.27

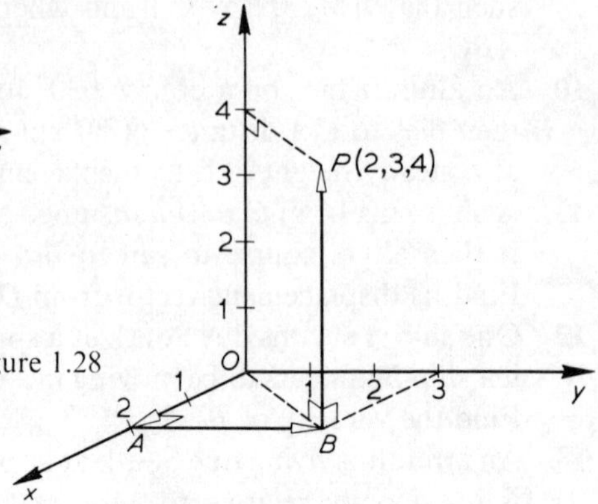

Figure 1.28

Any point in space has its own set of coordinates. For example in Figure 1.28 the point $P\,(2, 3, 4)$, defines the position which may be reached by a displacement of 2 units parallel to the x axis, **OA**, 3 units parallel to the y axis, **AB**, and then 4 units parallel to the z axis, **BP**.

Can you suggest other routes to P which are parallel to the axes? You will find that they are all included in the edges of the rectangular solid, $OABCMNPQ$ (Figure 1.29).

Figure 1.29

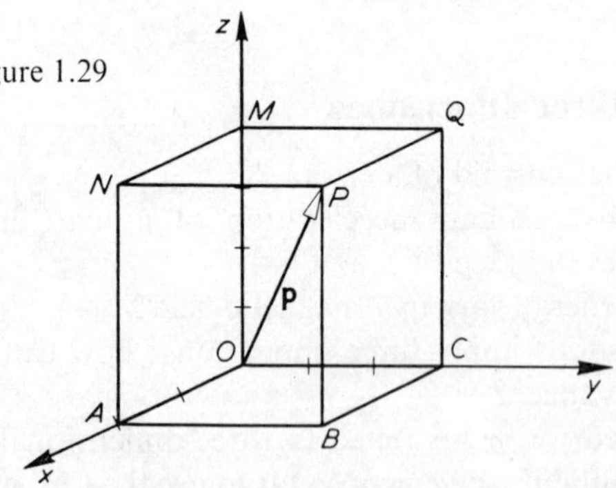

The resultant displacement is always the diagonal **OP** or $\mathbf{p} = [2, 3, 4]$.

The unit vectors **i**, **j** *and* **k**

So that we can use the symbols x, y and z as numbers, we now define vectors **i**, **j** and **k** as those vectors each of magnitude 1 unit and in the direction of the x, y and z axes, respectively.

Therefore

$$\mathbf{i} = [1, 0, 0], \; \mathbf{j} = [0, 1, 0] \text{ and } \mathbf{k} = [0, 0, 1].$$

A displacement of two units in the x direction only is

$$2\mathbf{i} = [2, 0, 0]$$

a displacement of three units in the y direction only is

$$3\mathbf{j} = [0, 3, 0]$$

and a displacement of 4 units in the z direction only is

$$4\mathbf{k} = [0, 0, 4]$$

Therefore we represent **p** by the vector $2\mathbf{i}+3\mathbf{j}+4\mathbf{k}$, where the addition sign has the same meaning as we gave it on page 13, or by the vector [2, 3, 4].

The numbers 2, 3, 4 are called the components or *projections* of **p** on to the three perpendicular axes.

What is a *projection*?

Figure 1.30

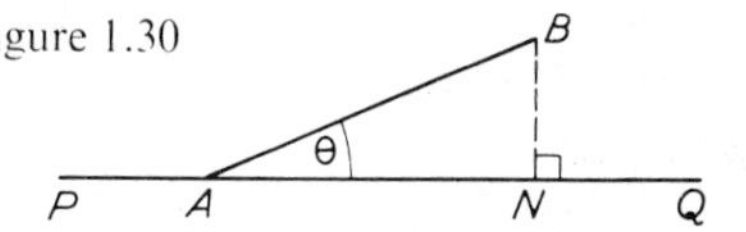

Suppose we project the segment AB on to the line PQ (Figure 1.30). We drop a perpendicular from B to PQ and the projection is then defined as AN. From the right-angled triangle ANB, we have

$$AN = AB \cos \theta$$

Figure 1.31

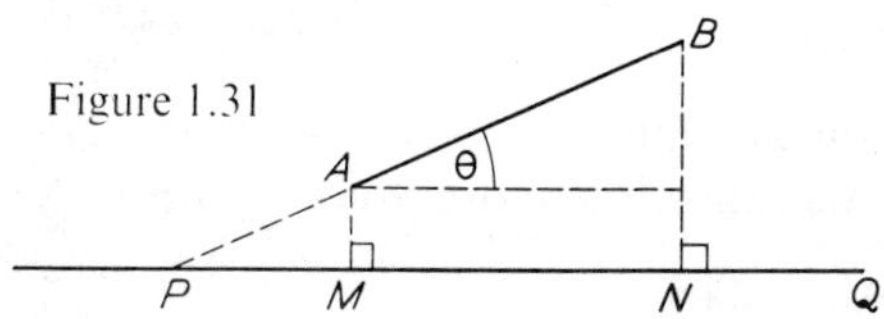

Similarly in Figure 1.31, to obtain the projection of AB on to PQ, we draw perpendiculars AM and BN to PQ. The distance between the feet of these perpendiculars is the projection of AB on to PQ, that is

$$MN = AB\cos\theta$$

Since a vector has both magnitude and direction, what are the magnitude and direction of $\mathbf{p}$?

Figure 1.32

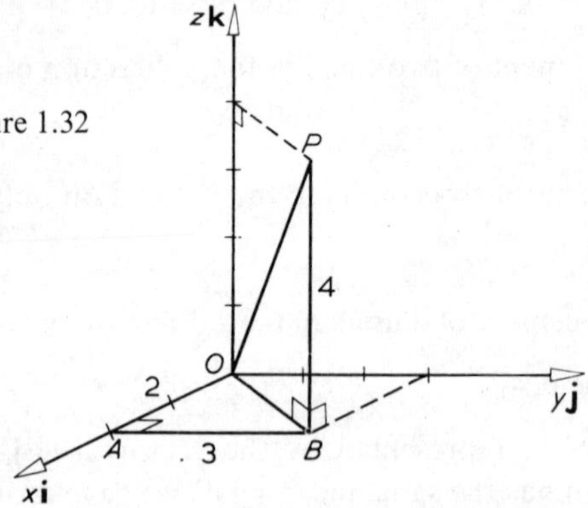

One way to find the magnitude of $\mathbf{p}$, written p, is to consider the right-angled triangles OAB and OBP (Figure 1.32).

By Pythagoras' theorem, we have

$$OB^2 = OA^2 + AB^2$$

and

$$\begin{aligned} OP^2 &= OB^2 + BP^2 \\ &= OA^2 + AB^2 + BP^2 \end{aligned}$$

Since OA, AB and BP are the components of $\mathbf{p}$ in the directions of $\mathbf{i}$, $\mathbf{j}$ and $\mathbf{k}$, respectively, we have

$$\begin{aligned} p^2 &= 4+9+16 \\ &= 29 \end{aligned}$$

Therefore the magnitude of $\mathbf{p}$ is $\sqrt{29}$ units.

In three dimensions, therefore, for vector $\mathbf{r}$ we have

$$\mathbf{r} = x\mathbf{i} + y\mathbf{j} + z\mathbf{k} = [x, y, z]$$

where

$$r^2 = x^2 + y^2 + z^2$$

or

$$r = |\mathbf{r}| = \sqrt{x^2+y^2+z^2}$$

The components x, y and z are also known as the *resolutes* or resolved parts of **r** on to the three mutually perpendicular axes in the directions of the unit vectors **i**, **j** and **k**, respectively.

Direction cosines

What is the direction of **p**?

There is a difficulty here because we have so far related direction to the x axis or perhaps the y axis. Now that we are in three dimensions, because one fixed angle alone cannot define direction in this space, we not only relate direction to both the x and y axes but also to the z axis.

Figure 1.33

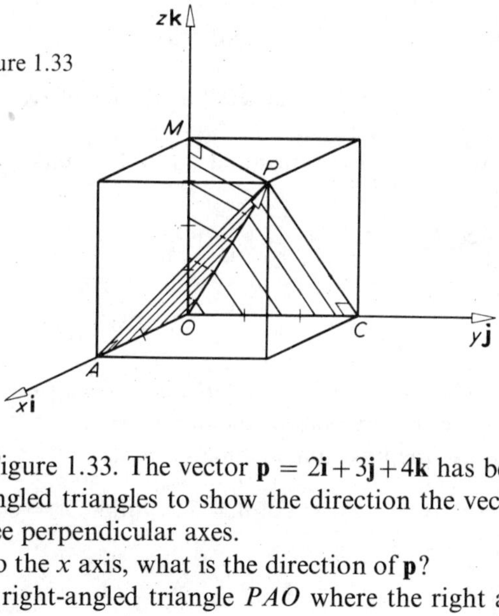

Look at Figure 1.33. The vector $\mathbf{p} = 2\mathbf{i}+3\mathbf{j}+4\mathbf{k}$ has been drawn with right-angled triangles to show the direction the vector makes with the three perpendicular axes.

Relative to the x axis, what is the direction of **p**?

From the right-angled triangle PAO where the right angle is at A, we have

$$\cos POA = \frac{AO}{OP} = \frac{2}{\sqrt{29}}$$

Therefore angle $POA = 68°\ 12'$.

Next, what is the direction of **p** relative to the y axis?

No doubt you have suggested the solution lies in the angle POC of the right-angled triangle PCO. Correct.

$$\cos POC = \frac{CO}{OP} = \frac{3}{\sqrt{29}}$$

Therefore angle $POC = 56° 9'$.

Finally from triangle PMO, we have

$$\cos POM = \frac{4}{\sqrt{29}}$$

Therefore angle $POM = 42° 2'$.

It is the cosine ratios that we use to define the direction of vectors.

If, therefore, for vector $\mathbf{r} = x\mathbf{i}+y\mathbf{j}+z\mathbf{k}$, α, β and γ are the angles which **r** makes with the axes in the direction of **i**, **j** and **k** respectively, then

$$\cos \alpha = x/r, \quad \cos \beta = y/r \quad \text{and} \quad \cos \gamma = z/r.$$

These ratios are called the direction cosines of **r**.

Are these ratios independent?

Suppose we square the ratios and add the results

$$\cos^2\alpha+\cos^2\beta+\cos^2\gamma = \frac{x^2}{r^2}+\frac{y^2}{r^2}+\frac{z^2}{r^2}$$

$$= \frac{x^2+y^2+z^2}{r^2} = \frac{r^2}{r^2} = 1$$

Therefore the ratios are not independent because we can obtain one of them from the values of the other two.

We often use the less cumbersome symbols

$$l = \cos \alpha = x/r, \quad m = \cos \beta = y/r \quad \text{and} \quad n = \cos \gamma = z/r$$

and then

$$l^2+m^2+n^2 = 1 \qquad \ldots \qquad \ldots \text{IV}$$

The unit vector $\hat{\mathbf{r}}$

We shall often require a vector of unit magnitude in the direction of a given vector. Let $\hat{\mathbf{r}}$ be of unit magnitude in the direction of **r**. What is the connection between $\hat{\mathbf{r}}$ and **r**?

From IV we have

$$\hat{\mathbf{r}} = l\mathbf{i}+m\mathbf{j}+n\mathbf{k}$$

Now

$$\mathbf{r} = x\mathbf{i}+y\mathbf{j}+z\mathbf{k}$$

$$= lr\mathbf{i}+mr\mathbf{j}+nr\mathbf{k} = r(l\mathbf{i}+m\mathbf{j}+n\mathbf{k})$$

that is

$$\mathbf{r} = r\hat{\mathbf{r}}$$

This gives an expression for $\hat{\mathbf{r}}$ in the form

$$\hat{\mathbf{r}} = \frac{\mathbf{r}}{r}$$

For $\mathbf{p} = 2\mathbf{i}+3\mathbf{j}+4\mathbf{k}$, we have

$$\hat{\mathbf{p}} = \frac{\mathbf{p}}{p} = \frac{2\mathbf{i}+3\mathbf{j}+4\mathbf{k}}{\sqrt{2^2+3^2+4^2}}$$

$$= \frac{1}{\sqrt{29}}(2\mathbf{i}+3\mathbf{j}+4\mathbf{k})$$

The direction cosines are

$$l = \frac{2}{\sqrt{29}}, \qquad m = \frac{3}{\sqrt{29}} \qquad \text{and} \qquad n = \frac{4}{\sqrt{29}}$$

From these values, we have

$$l^2+m^2+n^2 = \frac{4}{29}+\frac{9}{29}+\frac{16}{29} = \frac{29}{29}$$

$$= 1.$$

Having found l and m, we can always find n by using the relationship IV

$$n^2 = 1-l^2-m^2$$

Alternatively, we can use IV as confirmation of our calculation.

Example 1

If $\mathbf{a} = 5\mathbf{i}+2\mathbf{j}-\mathbf{k}$ and $\mathbf{b} = -3\mathbf{i}+4\mathbf{j}+10\mathbf{k}$, what is $\mathbf{a}+\mathbf{b}$?

We calculate resultants in three dimensions in a way similar to those in two. That is, we consider the components separately

$$\begin{aligned}\mathbf{a}+\mathbf{b} &= (5-3)\mathbf{i}+(2+4)\mathbf{j}+(-1+10)\mathbf{k}\\ &= 2\mathbf{i}+6\mathbf{j}+9\mathbf{k}\end{aligned}$$

Example 2

What is the magnitude of $\mathbf{a}+\mathbf{b}$?

$$\begin{aligned}|\mathbf{a}+\mathbf{b}|^2 &= 2^2+6^2+9^2\\ &= 4+36+81 = 121.\end{aligned}$$

Therefore $|\mathbf{a}+\mathbf{b}| = 11$.

Example 3

What is the unit vector in the direction of $\mathbf{a}+\mathbf{b}$?

From the results of the previous two calculations, we have the unit vector

$$\frac{\mathbf{a}+\mathbf{b}}{|\mathbf{a}+\mathbf{b}|} = \frac{1}{11}(2\mathbf{i}+6\mathbf{j}+9\mathbf{k})$$

Example 4

What are the direction cosines of $\mathbf{a}+\mathbf{b}$?

$$l = \frac{2}{11}, \qquad m = \frac{6}{11} \qquad \text{and} \qquad n = \frac{9}{11}$$

Example 5

If $\mathbf{c} = 2\mathbf{i}+4\mathbf{j}-6\mathbf{k}$ and $\mathbf{d} = 3\mathbf{i}+6\mathbf{j}-9\mathbf{k}$, what is the resultant of $3\mathbf{c}-2\mathbf{d}$?

$$\begin{aligned}3\mathbf{c}-2\mathbf{d} &= \{3(2)-2(3)\}\mathbf{i}+\{3(4)-2(6)\}\mathbf{j}+\{3(-6)-2(-9)\}\mathbf{k}\\ &= (6-6)\mathbf{i}+(12-12)\mathbf{j}+(-18+18)\mathbf{k}\\ &= 0\mathbf{i}+0\mathbf{j}+0\mathbf{k} = \mathbf{0}\end{aligned}$$

This is the zero vector.

What, therefore, can we say about the vectors $3\mathbf{c}$ and $2\mathbf{d}$?

They are equal.

Another way to obtain this result is

$$3\mathbf{c} = 3(2\mathbf{i}+4\mathbf{j}-6\mathbf{k}) = 6\mathbf{i}+12\mathbf{j}-18\mathbf{k} = [6, 12, -18]$$

$$2\mathbf{d} = 2(3\mathbf{i}+6\mathbf{j}-9\mathbf{k}) = 6\mathbf{i}+12\mathbf{j}-18\mathbf{k} = [6, 12, -18]$$

For two vectors to be equal in three dimensions, therefore, we have not only the same conditions present as for two dimensions but we must ensure that the compontents in the direction of **k**, that is parallel to the z axis, are also equal.

Hence, when p_x, p_y, p_z and q_x, q_y, q_z are the components of **p** and **q** in the **i**, **j** and **k** directions, respectively, if

$$\mathbf{p} = \mathbf{q}$$

then we have $\mathbf{p}-\mathbf{q} = \mathbf{0}$

that is $(p_x-q_x)\mathbf{i}+(p_y-q_y)\mathbf{j}+(p_z-q_z)\mathbf{k} = \mathbf{0}$

For the left hand side of this equation to be zero, the three resolutes must be independently zero, that is

$$p_x = q_x, \qquad p_y = q_y \qquad \text{and} \qquad p_z = q_z.$$

Summary

A vector in the Cartesian plane is a number pair $\mathbf{r} = [x, y]$.

The magnitude of vector **r** is given by $r = \sqrt{x^2+y^2}$.

The direction **r** makes with the positive direction of the x axis is given by $\tan\theta = y/x$ where θ is the angle between vector **r** and the x axis.

Vectors are equivalent only when they have the same magnitude and the same direction.

Vector addition obeys the parallelogram law: if $\mathbf{p} = [x_1, y_1]$, and $\mathbf{q} = [x_2, y_2]$, then

$$\mathbf{p}+\mathbf{q} = [(x_1+x_2), (y_1+y_2)]$$

The vector $m\mathbf{p}$ is m times the magnitude of **p** and in the direction of **p** when $m > 0$ and in the direction opposite to that of **p** when $m < 0$.

Vectors **p** and **q** are parallel when $\mathbf{p} = m\mathbf{q}$ for $m \neq 0$.

The zero vector **0** has no magnitude and no defined direction.

The inverse vector satisfies the equation $\mathbf{p}+(-\mathbf{p}) = \mathbf{0}$.

The centroid of n given points is

$$\mathbf{g} = \frac{\mathbf{a}_1 + \mathbf{a}_2 + \ldots + \mathbf{a}_n}{n}$$

where the $\mathbf{a}_i$ are the vectors of the given points.

The centre of mass of a given number of masses is

$$\mathbf{g} = \frac{m_1\mathbf{a}_1 + m_2\mathbf{a}_2 + \ldots + m_n\mathbf{a}_n}{m_1 + m_2 + \ldots + m_n}$$

where the m_i are the masses of the particles with corresponding position vectors $\mathbf{a}_i$.

The velocity of A relative to B is given by the relationship

$$_A\mathbf{v}_B = {_A\mathbf{v}_O} - {_B\mathbf{v}_O} = \mathbf{v}_A - \mathbf{v}_B$$

A vector in three dimensions is given by $\mathbf{r} = x\mathbf{i} + y\mathbf{j} + z\mathbf{k}$, where $\mathbf{i}$, $\mathbf{j}$ and $\mathbf{k}$ are unit vectors, that is, vectors of magnitude 1 unit, in directions parallel to the x, y and z axes, respectively.

Vector $\mathbf{r}$ can also be represented by the number triple $[x, y, z]$. In this case the unit vectors $\mathbf{i}$, $\mathbf{j}$ and $\mathbf{k}$ are [1, 0, 0], [0, 1, 0] and [0, 0, 1], respectively.

The magnitude of $\mathbf{r}$ is $r = \sqrt{x^2 + y^2 + z^2}$.

The direction cosines of $\mathbf{r}$ are

$$l = \cos\alpha = x/r, \quad m = \cos\beta = y/r \quad \text{and} \quad n = \cos\gamma = z/r.$$

When $\mathbf{p} = \mathbf{q}$, we must have

$$p_x = q_x, \qquad p_y = q_y, \qquad \text{and} \qquad p_z = q_z.$$

$\hat{\mathbf{r}}$ is the unit vector, that is a vector of magnitude 1 unit, in the direction of $\mathbf{r}$ and it is calculated by finding $\mathbf{r}/r$.

Vector addition obeys the following laws:

1. $\mathbf{p} + \mathbf{q} = \mathbf{q} + \mathbf{p}$	Commutative
2. $(\mathbf{p} + \mathbf{q}) + \mathbf{r} = \mathbf{p} + (\mathbf{q} + \mathbf{r})$	Associative
3. $\mathbf{p} + \mathbf{0} = \mathbf{p}$	Identity
4. $\mathbf{p} + (-\mathbf{p}) = \mathbf{0}$	Inverse
5. $(\lambda + \mu)\mathbf{p} = \lambda\mathbf{p} + \mu\mathbf{p}$	Distributive
6. $\lambda(\mathbf{p} + \mathbf{q}) = \lambda\mathbf{p} + \lambda\mathbf{q}$	Distributive
7. $(\lambda\mu)\mathbf{p} = \lambda(\mu\mathbf{p})$	Distributive
8. $1\mathbf{p} = \mathbf{p}$	Unit element
9. $0\mathbf{p} = \mathbf{0}$	Zero element

Exercise 6

1 If $\mathbf{p} = 4\mathbf{i}+3\mathbf{j}$ and $\mathbf{q} = \mathbf{i}-\mathbf{j}$, find
(i) $\mathbf{p}+\mathbf{q}$, (ii) $\mathbf{p}-\mathbf{q}$, (iii) $2\mathbf{p}+\mathbf{q}$, (iv) $\mathbf{p}-4\mathbf{q}$. (v) $3\mathbf{p}-5\mathbf{q}$

2 What are the magnitudes of $\mathbf{p}$ and $\mathbf{q}$?

3 What directions do $\mathbf{p}$ and $\mathbf{q}$ make with the positive x axis?

4 Show that $|\mathbf{i}+\mathbf{j}| = |\mathbf{i}-\mathbf{j}| = \sqrt{2}$ units.

5 Calculate (i) $|3\mathbf{i}+4\mathbf{j}|$, (ii) $|3\mathbf{i}-4\mathbf{j}|$, (iii) $|4\mathbf{i}-5\mathbf{j}|$, (iv) $|-3\mathbf{i}+2\mathbf{j}|$, (v) $|-2\mathbf{i}-\mathbf{j}|$

6 Calculate the directions of the vector resultants
(i) $\mathbf{i}+\mathbf{j}$ (ii) $\mathbf{i}-\mathbf{j}$ (iii) $-\mathbf{i}-\mathbf{j}$ (iv) $-\mathbf{i}+\mathbf{j}$
(v) $\mathbf{i}+2\mathbf{j}$ (vi) $-\mathbf{i}+2\mathbf{j}$ (vii) $2\mathbf{i}+3\mathbf{j}$ (viii) $2\mathbf{i}-3\mathbf{j}$
(ix) $-2\mathbf{i}+3\mathbf{j}$ (x) $-2\mathbf{i}-3\mathbf{j}$

7 If $\mathbf{a} = 2\mathbf{i}-\mathbf{j}+3\mathbf{k}$, $\mathbf{b} = \mathbf{i}+\mathbf{j}-2\mathbf{k}$ and $\mathbf{c} = -3\mathbf{i}-2\mathbf{j}+\mathbf{k}$, find
(i) $\mathbf{a}+2\mathbf{b}$ (ii) $2\mathbf{a}-\mathbf{b}$ (iii) $\mathbf{b}+3\mathbf{c}$
(iv) $\mathbf{a}+2\mathbf{b}+\mathbf{c}$ (v) $2\mathbf{a}-\mathbf{b}-\mathbf{c}$ (vi) $\mathbf{a}+\mathbf{b}+3\mathbf{c}$

8 What are the direction cosines of $\mathbf{a}$, $\mathbf{b}$ and $\mathbf{c}$?

9 Show that $|\mathbf{i}+\mathbf{j}+\mathbf{k}| = |\mathbf{i}-\mathbf{j}-\mathbf{k}| = |\mathbf{i}+\mathbf{j}-\mathbf{k}| = \sqrt{3}$ units.

10 Calculate (i) $|\mathbf{i}+2\mathbf{j}+2\mathbf{k}|$ (ii) $|2\mathbf{i}+\mathbf{j}+2\mathbf{k}|$
(iii) $|4\mathbf{i}+4\mathbf{j}+2\mathbf{k}|$ (iv) $|6\mathbf{i}-6\mathbf{j}-3\mathbf{k}|$ (v) $|\mathbf{i}+4\mathbf{j}-8\mathbf{k}|$

11 If the direction cosines of $\mathbf{i}+\mathbf{j}+\mathbf{k}$ are l, m and n, respectively, find l, m and n and show that $l^2+m^2+n^2 = 1$.

12 Find the direction cosines of (i) $\mathbf{i}+4\mathbf{j}+8\mathbf{k}$ (ii) $\mathbf{i}-4\mathbf{j}+8\mathbf{k}$
(iii) $4\mathbf{i}-4\mathbf{j}-7\mathbf{k}$ (iv) $\mathbf{i}+2\mathbf{j}+3\mathbf{k}$ (v) $-2\mathbf{i}+\mathbf{j}-2\mathbf{k}$

13 What are the unit vectors of question 12?

14 If the vectors of A and B are $\mathbf{i}+\mathbf{j}$ and $4\mathbf{i}+5\mathbf{j}$, respectively, find the vector $\mathbf{AB}$, its magnitude and direction.

15 Use the data of question 14 to find the vector $\mathbf{BA}$, its magnitude and direction.

16 If the vectors of M and N are $3\mathbf{i}-5\mathbf{j}$ and $-2\mathbf{i}+7\mathbf{j}$, respectively, find $\mathbf{MN}$, its magnitude and direction.

17 Use the data of question 16 to find $\mathbf{NM}$, its magnitude and direction.

18 If the vectors of P and Q are $4\mathbf{i}-7\mathbf{j}+5\mathbf{k}$ and $2\mathbf{i}-6\mathbf{j}-3\mathbf{k}$, respectively, find the vector $\mathbf{PQ}$, its magnitude and direction cosines.

19 Repeat question 18 for the vectors $3\mathbf{i}+\mathbf{j}-\mathbf{k}$ and $\mathbf{i}-2\mathbf{j}+\mathbf{k}$.

20 From the data of questions 18 and 19 find the vectors $\mathbf{QP}$, their magnitudes and directions.

21 If $\mathbf{a} = 3\mathbf{i}+4\mathbf{j}$ and $\mathbf{b} = \mathbf{i}-2\mathbf{j}+2\mathbf{k}$, what are the magnitudes and directions of the vectors $\hat{\mathbf{a}}$ and $\hat{\mathbf{b}}$?

22 Prove that the lines joining the mid-points of pairs of opposite edges of a tetrahedron bisect each other.

23 Let **a**, **b**, **c** and **d** be the vectors of the vertices A, B, C and D, respectively, of a tetrahedron. Prove that the centroid of the vertices of the tetrahedron is $\frac{1}{4}(\mathbf{a}+\mathbf{b}+\mathbf{c}+\mathbf{d})$.

24 Masses of 1, 2 and 3 kg are at points (1, 1, 4), (−2, 1, 4) and (1, −5, −6), respectively. Find the vector of the centre of mass.

25 What is the vector of the centre of mass of masses 4, 3 and 5 kg at the points (−1, 1, 5), (2, 0, −4) and (2, 4, −4), respectively?

26 What are the coordinates of the centre of mass of masses 1, 2, 3 and 4 kg at points (0, 1, 4), (2, 2, −3), (−2, 5, −2) and (3, 0, −3), respectively?

27 Two particles have the following position and velocity vectors measured at the same time in units of metres and metres second^{-1}, respectively.

$$\mathbf{s}_1 = 10\mathbf{i} \text{ and } \mathbf{v}_1 = 30\mathbf{i} \text{ and } \mathbf{s}_2 = -10\mathbf{i} \text{ and } \mathbf{v}_2 = 40\mathbf{i}$$

If the particles continue with the same velocities, find their position vectors after 1, 2 and 3 seconds.

When does the second particle overtake the first?

28 At noon two cyclists have the following position and velocity vectors

$$\mathbf{s}_1 = 15\mathbf{j} \text{ and } \mathbf{v}_1 = -10\mathbf{j} \text{ and } \mathbf{s}_2 = -20\mathbf{i} \text{ and } \mathbf{v}_2 = 15\mathbf{i}$$

where the units are metres and m s^{-1} respectively.

What is the distance between them at noon and at 1 and 2 seconds later, if they continue with the same velocities?

29 Two ships have the following position and velocity vectors at noon and the units are kilometres and km h^{-1}

$$\mathbf{s}_1 = \mathbf{i}+\mathbf{j} \quad \text{and} \quad \mathbf{v}_1 = 5\mathbf{i}+10\mathbf{j}$$
$$\mathbf{s}_2 = 2\mathbf{i}+3\mathbf{j} \quad \text{and} \quad \mathbf{v}_2 = 3\mathbf{i}-7\mathbf{j}$$

What are their respective position vectors at 2 p.m.? Find the distance between them at this time.

30 Particles A and B start simultaneously from O with constant velocities **i** and **j**, respectively. Find the velocity of A relative to B.

What are the magnitude and direction of this relative velocity? The units are 1 m s^{-1} for **i** and **j**, East and North respectively.

31 Particles A and B start simultaneously from O with constant velocity vectors $5\mathbf{i}-2\mathbf{j}$ and $9\mathbf{i}-14\mathbf{j}$, respectively. Find the velocity of B relative to A.

What are the magnitude and direction of this relative velocity?

32 The velocity of a particle A relative to a particle B is $2\mathbf{i}-3\mathbf{j}$, and the velocity of B relative to a third particle C is $\mathbf{i}-\mathbf{j}$. Find the relative velocity of A to C.
What are the magnitude and direction of this relative velocity?

33 The velocity of a ship relative to the sea is $6\mathbf{i}+7\mathbf{j}$ and that of the sea relative to the Earth is $2\mathbf{i}-\mathbf{j}$. Find the relative velocity of the ship to the Earth if $\mathbf{i}$ and $\mathbf{j}$ represent velocity vectors of 1 knot, East and North, respectively.

34 If $\mathbf{v}_A-\mathbf{v}_B=[-1, 3, -6]$ and $\mathbf{v}_B-\mathbf{v}_C=[1, -3, 6]$ find $\mathbf{v}_A$ and $\mathbf{v}_C$ if $\mathbf{v}_B=[2, -1, 4]$. What are the magnitudes of the velocities of A and C and determine any relationship that may exist between $\mathbf{v}_A$ and $\mathbf{v}_C$?

2 · *Vector multiplication*

Because vectors were originally invented to simplify solutions to physical problems, the results of their algebra must be consistent with what we find in science: it took many years to construct such an algebra.

We know that addition of vectors is performed in a special way. The same is true when two vectors are 'multiplied' together. In fact, there are two different methods of vector multiplication.

Scalar or dot product

We have already seen that the use of projections to describe a vector in space is fundamental. We now consider the projection of one line on another in space, which leads us to the first method of multiplication.

Suppose $\mathbf{p} = p_x\mathbf{i}+p_y\mathbf{j}+p_z\mathbf{k}$ and $\mathbf{q} = q_x\mathbf{i}+q_y\mathbf{j}+q_z\mathbf{k}$, and we find the projection of $\mathbf{p}$ on to $\mathbf{q}$, that is $\mathbf{ON}$ (Figure 2.1).

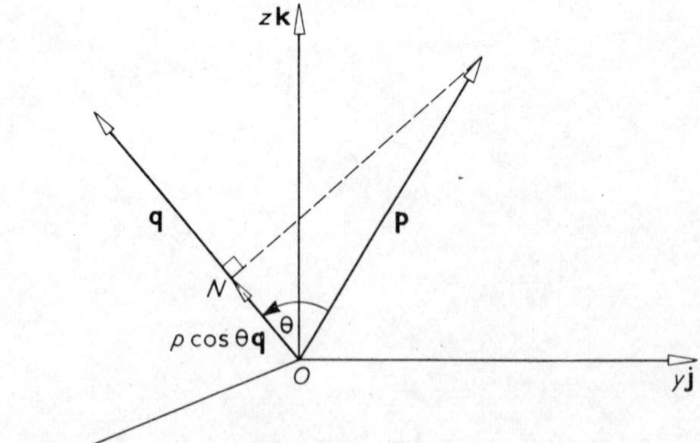

Figure 2.1

Both **p** and **q** are given in terms of their projections on to the three mutually orthogonal axes, and so we begin by finding the projection of the unit vector **i** on to **q** (Figure 2.2).

Figure 2.2

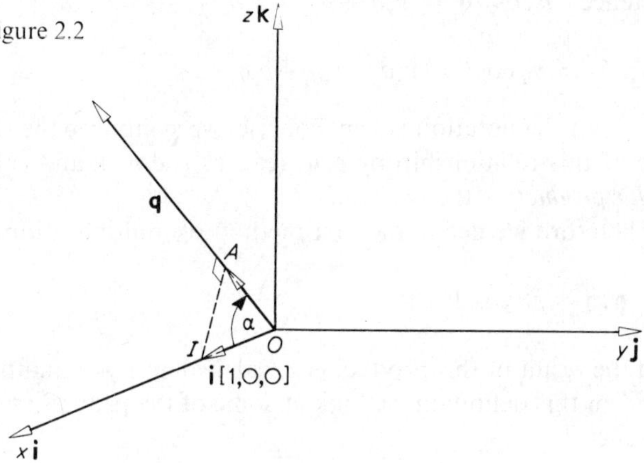

When we project **i** on to **q**, we obtain the vector **OA**

$$\mathbf{OA} = |\mathbf{i}| \cos\alpha\, \hat{\mathbf{q}}$$

where $\hat{\mathbf{q}}$ is the unit vector in the direction of **q**.

Now, $\cos\alpha = q_x/q$, where

$$|\mathbf{i}| \cos\alpha\, \hat{\mathbf{q}} = 1\frac{q_x}{q}\hat{\mathbf{q}} = q_x\frac{\hat{\mathbf{q}}}{q}$$

Therefore the projection of $p_x\mathbf{i}$ on to **q** is $p_xq_x\frac{\hat{\mathbf{q}}}{q}$.

Similarly the projections of $p_y\mathbf{j}$ and $p_z\mathbf{k}$ on to **q** are

$$p_y|\mathbf{j}| \cos\beta\, \hat{\mathbf{q}} = p_yq_y\frac{\hat{\mathbf{q}}}{q}$$

$$p_z|\mathbf{k}| \cos\gamma\, \hat{\mathbf{q}} = p_zq_z\frac{\hat{\mathbf{q}}}{q}$$

Therefore the projection of **p** on to **q** is

$$p_xq_x\frac{\hat{\mathbf{q}}}{q}+p_yq_y\frac{\hat{\mathbf{q}}}{q}+p_zq_z\frac{\hat{\mathbf{q}}}{q} = (p_xq_x+p_yq_y+p_zq_z)\frac{\hat{\mathbf{q}}}{q}$$

From Figure 2.1 we see, however, that the projection vector of **p** on to **q** is given by $p\cos\theta\,\hat{\mathbf{q}}$, where θ is the angle between the vectors.

Hence $$p\cos\theta\,\hat{\mathbf{q}} = (p_x q_x + p_y q_y + p_z q_z)\frac{\hat{\mathbf{q}}}{q},$$

or $$pq\cos\theta = p_x q_x + p_y q_y + p_z q_z$$

So that the notation is kept concise, we symbolize the right-hand side of this relationship by $\mathbf{p}.\mathbf{q}$, read as p dot q, and call this the *scalar product* or the *dot product*.

Therefore we define the first method of multiplication as

$$\mathbf{p}.\mathbf{q} = pq\cos\theta$$

and the result of this product is a scalar quantity, a number.

From this definition we look at some of the properties of the dot product.

1. $\cos\theta = \dfrac{\mathbf{p}.\mathbf{q}}{pq}$

Though the cosine given here is in terms of an arbitrary definition, $\mathbf{p}.\mathbf{q}$, it is the actual cosine of the angle between the two vectors. By the Cauchy-Schwarz inequality, which we shall not prove here, $|\mathbf{p}.\mathbf{q}| \leqslant |\mathbf{p}|\,|\mathbf{q}|$, and so $\dfrac{\mathbf{p}.\mathbf{q}}{pq}$ lies between the limits of -1 and $+1$.

We are, therefore, able to determine easily the angle between two vectors in space.

2. *Is* $\mathbf{p}.\mathbf{q} = \mathbf{q}.\mathbf{p}$?

Figure 2.3

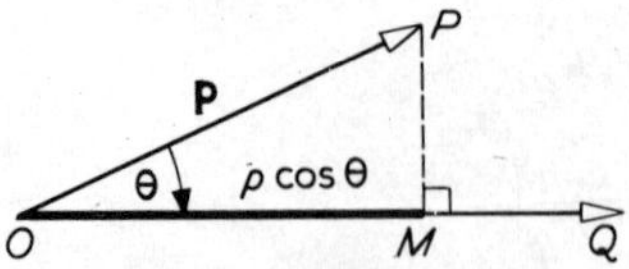

Figure 2.3 shows the projection of **p** on to **q**, that is

$$(p\cos\theta)q = \mathbf{OM}.\mathbf{OQ}$$

Figure 2.4

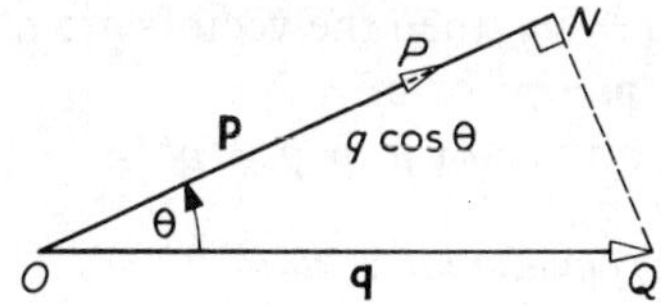

Figure 2.4 shows the projection of **q** on to **p**, that is

$(q\cos\theta)p = \mathbf{ON}.\mathbf{OP}$

Figure 2.5

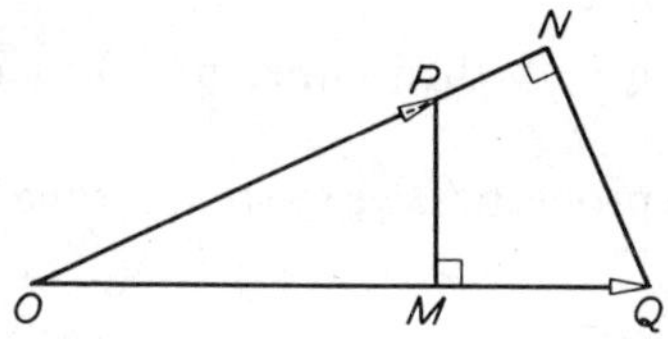

Figure 2.5 shows both cases, and because M and N are right angles, PQ is a diameter of a circle whose circumference passes through $PMQN$.

Therefore $\mathbf{OM}.\mathbf{OQ} = \mathbf{ON}.\mathbf{OP}$

that is $pq\cos\theta = qp\cos\theta$

Because the dot product, by definition, involves no direction, and the magnitudes $pq\cos\theta$ and $qp\cos\theta$ are equal, we have

$$\mathbf{p}.\mathbf{q} = \mathbf{q}.\mathbf{p}$$

An alternative way to show this is useful. Let us regard an anti-clockwise rotation as positive and a clockwise rotation as negative. Therefore the direction of θ from **q** to **p** on the diagrams is positive and the direction from **p** to **q** is negative. Now $\cos\theta = \cos(-\theta)$, and so the value of the dot product is unaltered when we alter the order of the letters. Hence

$$\mathbf{p}.\mathbf{q} = \mathbf{q}.\mathbf{p}$$

The dot product, therefore, is commutative.

3. *Dot product of parallel vectors*

When two vectors are parallel then $\theta = 0$ and $\cos\theta = 1$. Therefore

$$\mathbf{p}.\mathbf{q} = pq$$

Conversely, if $\mathbf{p}.\mathbf{q} = pq$, then the vectors are parallel. In particular, if $\mathbf{q} = \mathbf{p}$, then $\mathbf{p}.\mathbf{p} = p^2$ or $\mathbf{p}^2 = p^2$.

Further, for the unit vector $\hat{\mathbf{p}}$, $\hat{\mathbf{p}}.\hat{\mathbf{p}} = p^2 = 1$.

4. *Dot product of perpendicular vectors*

If the vectors are perpendicular, then $\theta = 90°$ and $\cos 90° = 0$. Therefore

$$\mathbf{p}.\mathbf{q} = 0$$

Conversely, if $\mathbf{p}.\mathbf{q} = 0$, then either $\mathbf{p} = 0$ or $\mathbf{q} = 0$ or $\mathbf{p}$ and $\mathbf{q}$ are perpendicular.

From these three properties, we see that for the three unit vectors, **i**, **j** and **k**, we have

$$\mathbf{i}.\mathbf{i} = \mathbf{i}^2 = 1$$

Similarly $\mathbf{j}.\mathbf{j} = \mathbf{k}.\mathbf{k} = 1$

Again $\mathbf{i}.\mathbf{j} = ij \cos 90° = 1.0 = 0$

Similarly $\mathbf{j}.\mathbf{k} = \mathbf{k}.\mathbf{i} = 0$

Summarising these results, we have

$$\mathbf{i}.\mathbf{i} = \mathbf{j}.\mathbf{j} = \mathbf{k}.\mathbf{k} = 1$$
$$\mathbf{i}.\mathbf{j} = \mathbf{j}.\mathbf{k} = \mathbf{k}.\mathbf{i} = 0$$

5. *Dot product in coordinate form*

Suppose $\mathbf{p} = p_x\mathbf{i}+p_y\mathbf{j}+p_z\mathbf{k}$ and $\mathbf{q} = q_x\mathbf{i}+q_y\mathbf{j}+q_z\mathbf{k}$: what is $\mathbf{p}.\mathbf{q}$?

$$\mathbf{p}.\mathbf{q} = (p_x\mathbf{i}+p_y\mathbf{j}+p_z\mathbf{k}).(q_x\mathbf{i}+q_y\mathbf{j}+q_z\mathbf{k}) = ?$$

How do we proceed?

One obvious way is to multiply the terms in the brackets in the same way as we do in number algebra. However, is this way valid for vector algebra?

In other words, is $\mathbf{a}.(\mathbf{b}+\mathbf{c}) = \mathbf{a}.\mathbf{b}+\mathbf{a}.\mathbf{c}$?

Notice first the different use of the + sign: on the left-hand side it is used to add two vectors together and on the right-hand side it is used to add together two numbers.

To prove this relationship, we say

$$\mathbf{a} = [a_x, a_y, a_z]$$

$$\mathbf{b}+\mathbf{c} = [(b_x+c_x), (b_y+c_y), (b_z+c_z)]$$

Hence $\quad \mathbf{a}.(\mathbf{b}+\mathbf{c}) = a_x(b_x+c_x)+a_y(b_y+c_y)+a_z(b_z+c_z)$

$$= a_xb_x+a_xc_x+a_yb_y+a_yc_y+a_zb_z+a_zc_z$$

and $\quad \mathbf{a}.\mathbf{b}+\mathbf{a}.\mathbf{c} = (a_xb_x+a_yb_y+a_zb_z)+(a_xc_x+a_yc_y+a_zc_z)$

$$= \mathbf{a}.(\mathbf{b}+\mathbf{c})$$

Therefore $\quad \mathbf{a}.(\mathbf{b}+\mathbf{c}) = \mathbf{a}.\mathbf{b}+\mathbf{a}.\mathbf{c}$

This is one of the distributive laws of the dot product.

Another distributive law which the student can prove in a similar way is the important relationship

$$\mathbf{a}.(m\mathbf{b}) = (m\mathbf{a}).\mathbf{b} = m(\mathbf{a}.\mathbf{b})$$

where m is a number.

Now we continue our example.

$$\begin{aligned}\mathbf{p}.\mathbf{q} &= (p_x\mathbf{i}+p_y\mathbf{j}+p_z\mathbf{k}).(q_x\mathbf{i}+q_y\mathbf{j}+q_z\mathbf{k})\\
&= p_x\mathbf{i}.q_x\mathbf{i}+p_x\mathbf{i}.q_y\mathbf{j}+p_x\mathbf{i}.q_z\mathbf{k}\\
&\quad +p_y\mathbf{j}.q_x\mathbf{i}+p_y\mathbf{j}.q_y\mathbf{j}+p_y\mathbf{j}.q_z\mathbf{k}\\
&\quad +p_z\mathbf{k}.q_x\mathbf{i}+p_z\mathbf{k}.q_y\mathbf{j}+p_z\mathbf{k}.q_z\mathbf{k}\\
&= p_xq_x\mathbf{i}^2+p_xq_y\mathbf{i}.\mathbf{j}+p_xq_z\mathbf{i}.\mathbf{k}\\
&\quad +p_yq_x\mathbf{j}.\mathbf{i}+p_yq_y\mathbf{j}^2+p_yq_z\mathbf{j}.\mathbf{k}\\
&\quad +p_zq_x\mathbf{k}.\mathbf{i}+p_zq_y\mathbf{k}.\mathbf{j}+p_zq_z\mathbf{k}^2\\
&= p_xq_x(1)+p_xq_y(0)+p_xq_z(0)\\
&\quad +p_yq_x(0)+p_yq_y(1)+p_yq_z(0)\\
&\quad +p_zq_z(0)+p_zq_y(0)+p_zq_z(1)\end{aligned}$$

Therefore $\quad \mathbf{p}.\mathbf{q} = p_xq_x+p_yq_y+p_zq_z$

Example 1

Let $\mathbf{p} = 3\mathbf{i}+\mathbf{j}-4\mathbf{k}$ and $\mathbf{q} = 4\mathbf{i}-3\mathbf{j}-\mathbf{k}$.

Calculate (i) $\mathbf{p}.\mathbf{q}$, (ii) the angle between the two vectors.

(i) $\mathbf{p}.\mathbf{q} = (3\mathbf{i}+\mathbf{j}-4\mathbf{k}).(4\mathbf{i}-3\mathbf{j}-\mathbf{k})$

$$= 12-3+4 = 13.$$

This is a scalar quantity.

(ii) By definition, $\mathbf{p}.\mathbf{q} = pq\cos\theta$ and this result we now know is 13.

$$|\mathbf{p}| = \sqrt{3^2+1^2+(-4)^2} \quad \text{and} \quad |\mathbf{q}| = \sqrt{4^2+(-3)^2+1^2}$$

that is $p = \sqrt{26}$, $q = \sqrt{26}$ and $pq = \sqrt{26}\sqrt{26} = 26$.

Since $pq \cos \theta = 13$,
we have, therefore,

$$\cos \theta = \frac{13}{26} = \frac{1}{2}$$

and $\quad \theta = 60°$.

Example 2

If the coordinates of the three vertices of a triangle are $A = (6, 3, 7)$, $B = (5, -4, 1)$ and $C = (-1, -3, 8)$, find the angle between the sides AB and BC.

Vertex A has vector $6\mathbf{i}+3\mathbf{j}+7\mathbf{k}$, or [6, 3, 7] and vertex B has vector [5, −4, 1].

Therefore, from page 25, we can say

$$\begin{aligned}\mathbf{AB} = \mathbf{b}-\mathbf{a} &= [(5-6), (-4-3), (1-7)] \\ &= [-1, -7, -6] \quad \text{or} \quad -\mathbf{i}-7\mathbf{j}-6\mathbf{k}\end{aligned}$$

Similarly

$$\begin{aligned}\mathbf{BC} = \mathbf{c}-\mathbf{b} &= [(-1-5), (-3-(-4), (8-1)] \\ &= [-6, 1, 7] \quad \text{or} \quad -6\mathbf{i}+\mathbf{j}+7\mathbf{k}.\end{aligned}$$

$$|\mathbf{AB}| = \sqrt{1+49+36} = \sqrt{86}$$

$$|\mathbf{BC}| = \sqrt{36+1+49} = \sqrt{86}$$

$$\mathbf{AB}.\mathbf{BC} = \sqrt{86}\sqrt{86}\cos \theta = 86 \cos \theta$$

Again $\mathbf{AB}.\mathbf{BC} = (-1, -7, -6).(-6, 1, 7) = -1(-6)-7(1)-6(7)$

$$= 6-7-42 = -43.$$

That is $\quad 86 \cos \theta = -43$

$$\cos \theta = -\frac{43}{86} = -\frac{1}{2}$$

Therefore $\quad \theta = 120°$

An application of the dot product in mechanics occurs in work. We define work done on a particle as the product of the force on the particle and the distance it is moved in the direction of the force. Work, like speed, is a scalar quantity and has no direction implied in its definition.

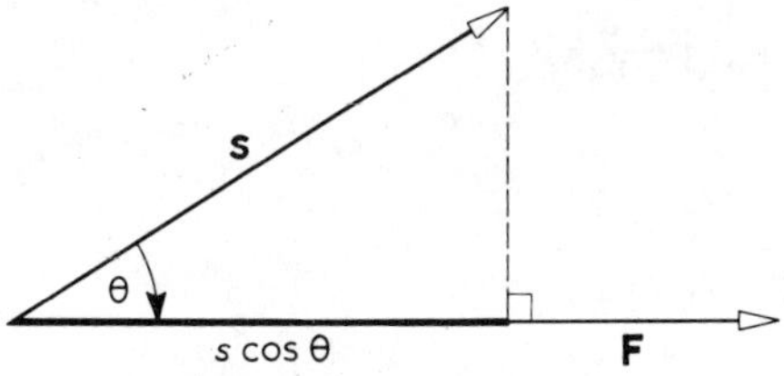

Figure 2.6

If, in Figure 2.6, we represent the force by the vector **F** and the displacement of the particle by **s**, then the work done, W, is

$$W = \mathbf{F}.\mathbf{s}$$
$$= Fs\cos\theta$$

where θ is the angle between the force and the displacement of the particle.

If $\mathbf{F} = F_x\mathbf{i}+F_y\mathbf{j}+F_z\mathbf{k}$ and $\mathbf{s} = s_x\mathbf{i}+s_y\mathbf{j}+s_z\mathbf{k}$, then

$$\mathbf{F}.\mathbf{s} = F_xs_x+F_ys_y+F_zs_z$$

Example 3

Two constant forces, 7 newtons and 5 newtons, act in the directions of vectors [4, 12, −3] and [−3, 4, 12] respectively. What is the work done on a particle which is displaced from $A = (4, -5, 0)$ to $B = (6, -2, -1)$?

The unit of length is 1 metre.

Let the first force be represented by the vector **p** and the second force by **q**. We know the magnitudes of the forces and now we require their direction. From page 36, we find the direction cosines of **p**

$$p_l = \frac{4}{13}, \qquad p_m = \frac{12}{13}, \qquad p_n = \frac{-3}{13}$$

Similarly the direction cosines of **q** are

$$q_l = \frac{-3}{13}, \qquad q_m = \frac{4}{13}, \qquad q_n = \frac{12}{13}$$

Since the forces have magnitudes 7 newtons and 5 newtons respectively, and we now know the coefficients of each unit vector

$$\mathbf{p} = 7\left(\frac{4}{13}\mathbf{i}+\frac{12}{13}\mathbf{j}-\frac{3}{13}\mathbf{k}\right)$$

$$= \frac{7}{13}(4\mathbf{i}+12\mathbf{j}-3\mathbf{k})$$

$$\mathbf{q} = 5\left(\frac{-3}{13}\mathbf{i}+\frac{4}{13}\mathbf{j}+\frac{12}{13}\mathbf{k}\right)$$

$$= \frac{5}{13}(-3\mathbf{i}+4\mathbf{j}+12\mathbf{k})$$

The resultant of $\mathbf{p}+\mathbf{q} = \mathbf{F}$, say

$$\mathbf{F} = \frac{1}{13}\{(28-15)\mathbf{i}+(84+20)\mathbf{j}+(-21+60)\mathbf{k}\}$$

$$= \frac{1}{13}\{13\mathbf{i}+104\mathbf{j}+39\mathbf{k}\}$$

$$= \mathbf{i}+8\mathbf{j}+3\mathbf{k}$$

Let the displacement $\mathbf{AB} = \mathbf{b}-\mathbf{a} = \mathbf{s}$, then

$$\mathbf{s} = [(6-4), (-2+5), (-1-0)]$$
$$= [2, 3, -1] = 2\mathbf{i}+3\mathbf{j}-\mathbf{k}$$

Therefore $\mathbf{F}.\mathbf{s} = (\mathbf{i}+8\mathbf{j}+3\mathbf{k}).(2\mathbf{i}+3\mathbf{j}-\mathbf{k})$

$$= (2+24-3)\text{ J} = 23 \text{ joules}$$

Example 4

We have seen that when $\mathbf{p}.\mathbf{q} = 0$, then the vectors are perpendicular if one or the other is non-zero. We use this property to show that the diagonals of a rhombus intersect at right angles.

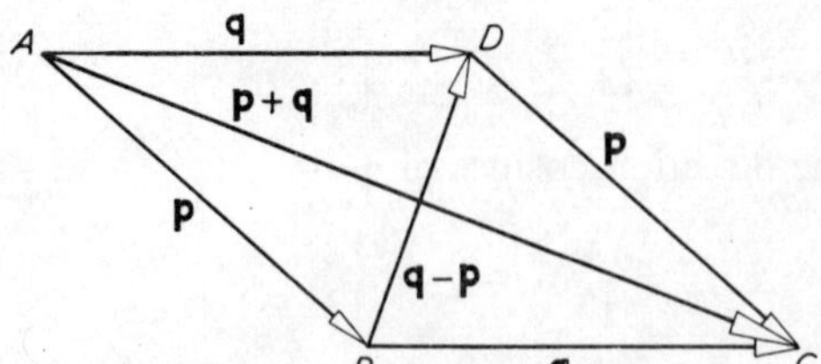

Figure 2.7

In Figure 2.7, **AB** = **p** and **AD** = **q**.
Therefore **AC** = **p**+**q** and **BD** = **q**−**p**.
Now **AC**.**BD** = (**p**+**q**).(**q**−**p**)

$$= \mathbf{p}.\mathbf{q}-\mathbf{p}^2+\mathbf{q}^2-\mathbf{q}.\mathbf{p}$$

Since **q**.**p** = **p**.**q** and in a rhombus |**p**| = |**q**|, so that $p^2 = q^2$ and $\mathbf{p}^2 = \mathbf{q}^2$, we have, therefore,

$$\mathbf{AC}.\mathbf{BD} = 0.$$

Hence the dot product is zero, and because **p** and **q** are non-zero vectors, the diagonals AC and BD are orthogonal.

Exercise 7

1 The vertices of a triangle are $A(0, 4)$, $B(1, 1)$ and $C(3, 3)$. N is the midpoint of BC. Find vectors **ON** and **AN** and show that ON is perpendicular to AN by
(*a*) the dot product,
(*b*) Pythagoras' theorem.

2 If $\mathbf{a} = [a_1, a_2]$ and $\mathbf{b} = [b_1, b_2]$, show that if **a**, **b** and **a**+**b** form a right-angled triangle, then $a_1b_1+a_2b_2 = 0$.

3 Obtain a similar equation when the directions of **a** and **b** are perpendicular to each other in the right-angled triangle formed by **a**, **b** and **a**+**b** where

$$\mathbf{a} = [a_1, a_2, a_3] \quad \text{and} \quad \mathbf{b} = [b_1, b_2, b_3].$$

4 Prove that the median joining the vertex and base of an isosceles triangle is perpendicular to the base.

5 Prove that the angle in a semi-circle is a right-angle.

6 Prove that the altitudes of a triangle are concurrent.

7 Find the work done by a force of 4 newtons applied at an angle of 60° to the direction of motion of an object which is moved through 10 metres.

8 Find the work done by a force of 5 newtons acting in the direction of the vector 3**i**+4**j** on a particle displaced from the origin 5 metres along the positive x axis.

In the next five questions let the unit of distance be the metre.

9 Find the work done by the force in question 8 if it displaces a particle from the origin to A whose position vector is 4**i**−**j**.

10 Find the work done by the force in question 8 if it displaces a particle from $A(3, -2)$ to $B(5, 3)$.

11 Find the work done by a force of 6 newtons acting in the direction of the vector $[2, -1, 2]$ on a particle displaced from the origin to the point $(4, -3, 1)$.

12 Find the work done by the force in question 11 if it displaces a particle from $(3, 1, -1)$ to $(4, -3, 1)$.

13 Two forces, one of 9 newtons in the direction of the vector $[1, 4, 8]$ and another of $\sqrt{24}$ newtons in the direction $[1, -1, -2]$, act on a particle displaced from $A(3, 2, 6)$ to $B(5, -1, 7)$. Find the work done on the particle.

14 If $\mathbf{a} = \mathbf{i}+\mathbf{j}+\mathbf{k}$ and $\mathbf{b} = 2\mathbf{i}-\mathbf{j}-2\mathbf{k}$, find (i) $\mathbf{a}.\mathbf{b}$, (ii) $\hat{\mathbf{a}}$ and $\hat{\mathbf{b}}$, (iii) the components and direction cosines of $\mathbf{a}$ and $\mathbf{b}$.

15 If $\mathbf{p} = -\mathbf{i}+2\mathbf{j}-3\mathbf{k}$, $\mathbf{q} = 3\mathbf{i}+\mathbf{j}-2\mathbf{k}$ and $\mathbf{r} = 2\mathbf{i}-3\mathbf{j}+\mathbf{k}$, find $\mathbf{p}.(\mathbf{q}+\mathbf{r})$ and $\mathbf{p}.\mathbf{q}+\mathbf{p}.\mathbf{r}$ and verify that

$$\mathbf{p}.(\mathbf{q}+\mathbf{r}) = \mathbf{p}.\mathbf{q}+\mathbf{p}.\mathbf{r}$$

16 Use the vectors of question 15 to find (i) $(\mathbf{p}.\mathbf{q})\mathbf{r}$ (ii) $\mathbf{p}(\mathbf{q}.\mathbf{r})$ (iii) $(\mathbf{p}.\mathbf{q})\mathbf{r}+\mathbf{p}(\mathbf{q}.\mathbf{r})$.

17 What is the angle between the vectors $\mathbf{a}$ and $\mathbf{b}$ if $\mathbf{a} = \mathbf{i}+\mathbf{j}+\mathbf{k}$ and $\mathbf{b} = \mathbf{i}-\mathbf{j}+\mathbf{k}$?

18 What are the angles of the triangle ABC where $A(-2, 0)$, $B(1, -1)$ and $C(2, 2)$?

19 Prove that in any triangle ABC, $\mathbf{a}^2 = \mathbf{b}^2+\mathbf{c}^2+2\mathbf{b}.\mathbf{c}$.

20 Let $\mathbf{a} = (\cos\theta)\mathbf{i}+(\sin\theta)\mathbf{j}$ and $\mathbf{b} = (\cos\phi)\mathbf{i}+(\sin\phi)\mathbf{j}$. Find $\mathbf{a}.\mathbf{b}$ and show that $\cos(\theta-\phi) = \cos\theta\cos\phi+\sin\theta\sin\phi$.

21 If $\mathbf{p}.(\mathbf{q}-\mathbf{r}) = 0$, is $\mathbf{q}$ always equal to $\mathbf{r}$? From your answer, deduce whether a vector cancellation law exists.

Vector or cross product

When we study the physical properties of electromagnetism and hydromechanics, we frequently find that two vector quantities operating in one plane have an effect in a plane perpendicular to them both. Therefore the second method of vector multiplication is initiated by the desire to find a vector $\mathbf{r} = x\mathbf{i}+y\mathbf{j}+z\mathbf{k}$ which is perpendicular to both $\mathbf{p} = p_x\mathbf{i}+p_y\mathbf{j}+p_z\mathbf{k}$ and $\mathbf{q} = q_x\mathbf{i}+q_y\mathbf{j}+q_z\mathbf{k}$.

We know from the dot product that for perpendicularity, we have

$$\mathbf{p}.\mathbf{r} = 0 \quad \text{and} \quad \mathbf{q}.\mathbf{r} = 0$$

that is $p_x x + p_y y + p_z z = 0$ 1

and $q_x x + q_y y + q_z z = 0$ 2

This pair of equations in three unknowns has no unique solution. Having found one solution set, we can find another by any scalar multiple of the first. Therefore if **r** is perpendicular to both **p** and **q**, so is a vector collinear with **r**.

To solve these equations we must avoid dividing by a coefficient because any one of them may be zero. We eliminate z, therefore, by multiplying equation 1 by q_z, equation 2 by p_z and then subtracting the subsequent equations.

Here is the result:

$$(p_x q_z - q_x p_z)x + (p_y q_z - q_y p_z)y = 0 \qquad \ldots \qquad \ldots 3$$

Since we cannot divide, let us arbitrarily set

$$x = p_y q_z - q_y p_z$$

and $y = -(p_x q_z - q_x p_z)$

so that equation 3 is satisfied.

When we substitute these values back into the original equations and simplify, we obtain

$$z = p_x q_y - q_x p_y$$

Therefore $\mathbf{r} = (p_y q_z - q_y p_z)\mathbf{i} - (p_x q_z - q_x p_z)\mathbf{j} + (p_x q_y - q_x p_y)\mathbf{k}$

and this vector is perpendicular to both **p** and **q**.

This is the meaning given to the 'product' of two directed line segments. It is known as the *vector product* or *cross product* and is written $\mathbf{p}_\times \mathbf{q}$, read as p cross q.

It is easier to remember this definition as a symbolic determinant

$$\mathbf{p}_\times \mathbf{q} = \begin{vmatrix} \mathbf{i} & \mathbf{j} & \mathbf{k} \\ p_x & p_y & p_z \\ q_x & q_y & q_z \end{vmatrix}$$

It is symbolic because the elements of the first row are vectors and we cannot evaluate such a 'determinant'. However, for our purposes we can expand this symbolic determinant on any row or column in the same way as an ordinary determinant, and because we can do this, the form is a useful aid to the memory.

Suppose $\mathbf{p} = [p_x, 0, 0]$ and $\mathbf{q} = [q_x, q_y, 0]$ (Figure 2.8). These are two vectors in the $\mathbf{i}$, $\mathbf{j}$ plane and the angle between them is θ.

$$\mathbf{p}_\times\mathbf{q} = \begin{vmatrix} \mathbf{i} & \mathbf{j} & \mathbf{k} \\ p_x & 0 & 0 \\ q_x & q_y & 0 \end{vmatrix} = p_x q_y \mathbf{k}$$

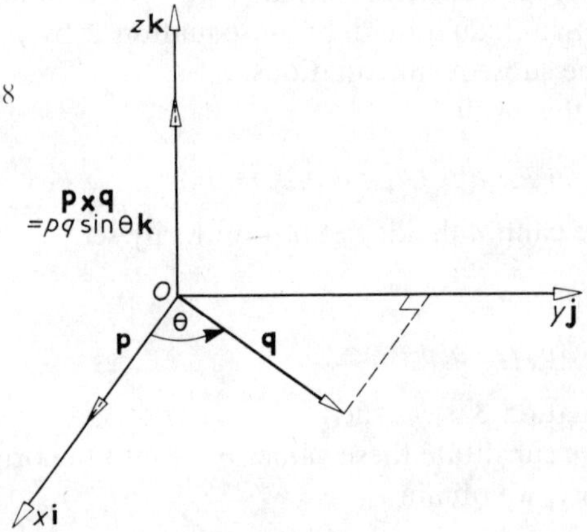

Figure 2.8

Now from Figure 2.8 we see that

$$p_x = p \quad \text{and} \quad q_y = q \sin\theta$$

Therefore $\mathbf{p}_\times\mathbf{q} = pq \sin\theta\, \mathbf{k}$ and $|\mathbf{p}_\times\mathbf{q}| = pq \sin\theta$

This result gives the magnitude of the cross product as the area of the parallelogram determined by $\mathbf{p}$ and $\mathbf{q}$. It leads us to another expression for the cross product

$$\mathbf{p}_\times\mathbf{q} = pq \sin\theta\, \hat{\mathbf{n}}$$

θ is the angle between $\mathbf{p}$ and $\mathbf{q}$. $\hat{\mathbf{n}}$ is a unit vector in the direction perpendicular to the plane containing $\mathbf{p}$ and $\mathbf{q}$ and in the sense of the motion of a right-handed screw turning from $\mathbf{p}$ through the smaller angle to $\mathbf{q}$ (Figure 2.9).

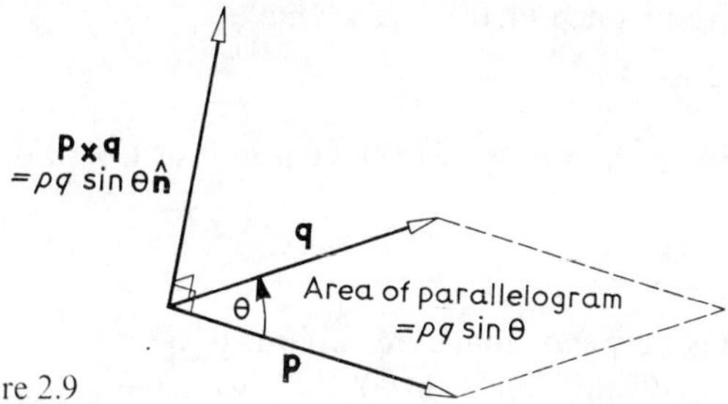

Figure 2.9

Notice again that the result of the cross product is itself a vector. From the two definitions

$$\mathbf{p}_{\times}\mathbf{q} = \begin{vmatrix} \mathbf{i} & \mathbf{j} & \mathbf{k} \\ p_x & p_y & p_z \\ q_x & q_y & q_z \end{vmatrix} = pq\sin\theta\,\hat{\mathbf{n}}$$

we next look at some of the properties of the cross product. Though these can be associated with the properties of determinants, we shall consider them using the definition $\mathbf{p}_{\times}\mathbf{q} = pq\sin\theta\,\hat{\mathbf{n}}$.

1. *Is* $\mathbf{p}_{\times}\mathbf{q} = \mathbf{q}_{\times}\mathbf{p}$?

Consider $\mathbf{q}_{\times}\mathbf{p}$. The magnitude of the resulting vector is

$$|\mathbf{q}_{\times}\mathbf{p}| = qp\sin\theta$$

What is the direction? This time we travel from **q** to **p** and since $\sin(-\theta) = -\sin\theta$, the direction is reversed. Hence

$$\mathbf{q}_{\times}\mathbf{p} = -pq\sin\theta\,\hat{\mathbf{n}}$$

This can also be seen from Figure 2.8: the motion of a right-handed screw from **q** to **p** is downwards from a horizontal plane.

Therefore the cross product is non-commutative

$$\mathbf{p}_{\times}\mathbf{q} = -\mathbf{q}_{\times}\mathbf{p}$$

2. *Cross product of parallel vectors*

Suppose **p** is parallel to **q**: what is $\mathbf{p}_{\times}\mathbf{q}$?

$$\mathbf{p}_{\times}\mathbf{q} = pq\sin\theta\,\hat{\mathbf{n}}$$

but $\theta = 0°$ and since $\sin 0° = 0$, we have

$$\mathbf{p}_{\times}\mathbf{q} = \mathbf{0}$$

Hence for $\mathbf{p}_{\times}\mathbf{q} = \mathbf{0}$, we can have $\mathbf{p} = \mathbf{0}$ or $\mathbf{q} = \mathbf{0}$ or $\mathbf{p}$ and $\mathbf{q}$ are parallel.

3. *Cross product of perpendicular vectors*

When $\mathbf{p}$ is perpendicular to $\mathbf{q}$, what is $\mathbf{p}_{\times}\mathbf{q}$?
Now $\theta = 90°$ and since $\sin 90° = 1$, we have

$$\mathbf{p}_{\times}\mathbf{q} = pq\,\hat{\mathbf{n}}$$

What does this mean? The three vectors, $\mathbf{p}$, $\mathbf{q}$ and $pq\hat{\mathbf{n}}$, form a set of three mutually perpendicular vectors.

We now apply these properties to the three unit vectors, $\mathbf{i}$, $\mathbf{j}$ and $\mathbf{k}$.

$$\mathbf{i}_{\times}\mathbf{i} = \mathbf{i}^2 \sin 0° \,\hat{\mathbf{n}} = \mathbf{0}$$

Similarly $\mathbf{j}_{\times}\mathbf{j} = \mathbf{k}_{\times}\mathbf{k} = \mathbf{0}$

Again $\mathbf{i}_{\times}\mathbf{j} = ij \sin 90° \,\hat{\mathbf{n}} = 1\hat{\mathbf{n}}$ in a direction perpendicular to both $\mathbf{i}$ and $\mathbf{j}$. Hence

$$\mathbf{i}_{\times}\mathbf{j} = \mathbf{k}$$

Similarly $\mathbf{j}_{\times}\mathbf{k} = \mathbf{i}$ and $\mathbf{k}_{\times}\mathbf{i} = \mathbf{j}$

The order of the multiplication is important, as we have seen, because the cross product is non-commutative. We have

$$\mathbf{j}_{\times}\mathbf{i} = -\mathbf{k}, \quad \mathbf{k}_{\times}\mathbf{j} = -\mathbf{i} \quad \text{and} \quad \mathbf{i}_{\times}\mathbf{k} = -\mathbf{j}.$$

These results are more conveniently stated as

$$\begin{aligned} \mathbf{i}_{\times}\mathbf{i} &= \mathbf{j}_{\times}\mathbf{j} = \mathbf{k}_{\times}\mathbf{k} = \mathbf{0} \\ \mathbf{i}_{\times}\mathbf{j} &= -\mathbf{j}_{\times}\mathbf{i} = \mathbf{k} \\ \mathbf{j}_{\times}\mathbf{k} &= -\mathbf{k}_{\times}\mathbf{j} = \mathbf{i} \\ \mathbf{k}_{\times}\mathbf{i} &= -\mathbf{i}_{\times}\mathbf{k} = \mathbf{j} \end{aligned}$$

4. *Cross product in cartesian form*

Let $\mathbf{p} = p_x\mathbf{i}+p_y\mathbf{j}+p_z\mathbf{k}$ and $\mathbf{q} = q_x\mathbf{i}+q_y\mathbf{j}+q_z\mathbf{k}$

then $\mathbf{p}_{\times}\mathbf{q} = (p_x\mathbf{i}+p_y\mathbf{j}+p_z\mathbf{k})_{\times}(q_x\mathbf{i}+q_y\mathbf{j}+q_z\mathbf{k}) = \,?$

We now ask the same question as at the corresponding stage in the dot product: how do we proceed?

Provided we remember the non-commutative property of the cross product, can we multiply the terms in the brackets in the same way as in number algebra?

Is $\mathbf{a}_\times(\mathbf{b}+\mathbf{c}) = \mathbf{a}_\times\mathbf{b}+\mathbf{a}_\times\mathbf{c}$?

Figure 2.10

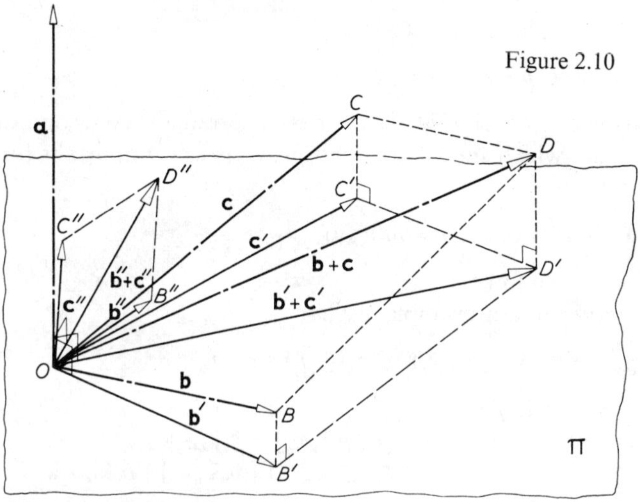

Figure 2.10 shows three vectors **a**, **b** and **c**. **a** is perpendicular to the plane Π whilst **b** and **c** are each inclined at different angles to Π. Perpendiculars are dropped from B, C and D to meet Π in B', C' and D' respectively.

Let $\mathbf{b}'$, $\mathbf{c}'$ and $\mathbf{b}'+\mathbf{c}'$ represent $\mathbf{OB}'$, $\mathbf{OC}'$ and $\mathbf{OD}'$, respectively. Further, we let $\mathbf{b}+\mathbf{c} = \mathbf{d}$ and $\mathbf{b}'+\mathbf{c}' = \mathbf{d}'$.

Now $\mathbf{a}_\times\mathbf{b} = ab\sin\theta\hat{\mathbf{n}}_1$, where θ is the angle between **a** and **b**, and $\mathbf{a}_\times\mathbf{b}' = ab'\sin 90^\circ\,\hat{\mathbf{n}}_1 = ab'\hat{\mathbf{n}}_1 = ab\cos(90^\circ-\theta)\hat{\mathbf{n}}_1 = ab\sin\theta\hat{\mathbf{n}}_1$.

Therefore $\mathbf{a}_\times\mathbf{b} = \mathbf{a}_\times\mathbf{b}' = ab'\hat{\mathbf{n}}_1$.

Similarly $\mathbf{a}_\times\mathbf{c} = \mathbf{a}_\times\mathbf{c}' = ac'\hat{\mathbf{n}}_2$

and $\mathbf{a}_\times\mathbf{d} = \mathbf{a}_\times\mathbf{d}' = ad'\hat{\mathbf{n}}_3$.

The vectors $ab'\hat{\mathbf{n}}_1$, $ac'\hat{\mathbf{n}}_2$ and $ad'\hat{\mathbf{n}}_3$ are all in the plane Π. They represent $\mathbf{OB}''$, $\mathbf{OC}''$ and $\mathbf{OD}''$ respectively, where $OB''D''C''$ are vertices of a quadrilateral whose sides are proportional to the sides of the parallelogram $OB'D'C'$.

Because the vectors $ab'\hat{\mathbf{n}}_1$, $ac'\hat{\mathbf{n}}_2$ and $ad'\hat{\mathbf{n}}_3$ have each been rotated through a right angle in the plane Π to obtain the vectors $\mathbf{OB}''$, $\mathbf{OC}''$ and $\mathbf{OD}''$, the angles of the quadrilateral $OB''D''C''$ are equal to the parallelogram $OB'D'C'$. In Figure 2.10, therefore, we have let these vectors be $\mathbf{b}''$, $\mathbf{c}''$ and $\mathbf{b}''+\mathbf{c}''$.

Therefore $OB''D''C''$ is a parallelogram and

$$\mathbf{OB''} + \mathbf{OC''} = \mathbf{OD''}$$

that is $ab'\,\hat{\mathbf{n}}_1 + ac'\,\hat{\mathbf{n}}_2 = ad'\,\hat{\mathbf{n}}_3$

or $\mathbf{a}_\times\mathbf{b} + \mathbf{a}_\times\mathbf{c} = \mathbf{a}_\times\mathbf{d}$

whence $\mathbf{a}_\times\mathbf{b} + \mathbf{a}_\times\mathbf{c} = \mathbf{a}_\times(\mathbf{b}+\mathbf{c})$

This is one distributive law for cross products. A second distributive law which the student can prove is also important for our work: it is

$$\mathbf{a}_\times(m\mathbf{b}) = (m\mathbf{a})_\times\mathbf{b} = m(\mathbf{a}_\times\mathbf{b})$$

where m is a number.

To return to our example, we have

$$\begin{aligned}
\mathbf{p}_\times\mathbf{q} &= (p_x\mathbf{i}+p_y\mathbf{j}+p_z\mathbf{k})_\times(q_x\mathbf{i}+q_y\mathbf{j}+q_z\mathbf{k}) \\
&= p_x\mathbf{i}_\times q_x\mathbf{i}+p_x\mathbf{i}_\times q_y\mathbf{j}+p_x\mathbf{i}_\times q_z\mathbf{k} \\
&\qquad +p_y\mathbf{j}_\times q_x\mathbf{i}+p_y\mathbf{j}_\times q_y\mathbf{j}+p_y\mathbf{j}_\times q_z\mathbf{k} \\
&\qquad\qquad +p_z\mathbf{k}_\times q_x\mathbf{i}+p_z\mathbf{k}_\times q_y\mathbf{j}+p_z\mathbf{k}_\times q_z\mathbf{k} \\
&= p_xq_x\mathbf{i}_\times\mathbf{i}+p_xq_y\mathbf{i}_\times\mathbf{j}+p_xq_z\mathbf{i}_\times\mathbf{k} \\
&\qquad +p_yq_x\mathbf{j}_\times\mathbf{i}+p_yq_y\mathbf{j}_\times\mathbf{j}+p_yq_z\mathbf{j}_\times\mathbf{k} \\
&\qquad\qquad +p_zq_x\mathbf{k}_\times\mathbf{i}+p_zq_y\mathbf{k}_\times\mathbf{j}+p_zq_z\mathbf{k}_\times\mathbf{k} \\
&= \mathbf{0}+p_xq_y\mathbf{k}-p_xq_z\mathbf{j} \\
&\qquad -p_yq_x\mathbf{k}+\ \mathbf{0}\ +p_yq_z\mathbf{i} \\
&\qquad\qquad +p_zq_x\mathbf{j}-p_zq_y\mathbf{i}+\mathbf{0} \\
&= (p_yq_z-p_zq_y)\mathbf{i}+(p_zq_x-p_xq_z)\mathbf{j}+(p_xq_y-p_yq_x)\mathbf{k}
\end{aligned}$$

This result leads us back to the symbolic determinant

$$\mathbf{p}_\times\mathbf{q} = \begin{vmatrix} \mathbf{i} & \mathbf{j} & \mathbf{k} \\ p_x & p_y & p_z \\ q_x & q_y & q_z \end{vmatrix}$$

Example 1

Find a unit vector perpendicular to vectors $\mathbf{p} = 6\mathbf{i}+5\mathbf{j}+2\mathbf{z}$ and $\mathbf{q} = 3\mathbf{i}+4\mathbf{j}+2\mathbf{k}$.

We first find the cross product which, though giving us a magnitude $|\mathbf{p}_\times\mathbf{q}|$ times the required unit magnitude, gives us the required direction.

$$\mathbf{p}_\times\mathbf{q} = \begin{vmatrix} \mathbf{i} & \mathbf{j} & \mathbf{k} \\ 6 & 5 & 2 \\ 3 & 4 & 2 \end{vmatrix}$$

$$= (10-8)\mathbf{i}-(12-6)\mathbf{j}+(24-15)\mathbf{k}$$

$$= 2\mathbf{i}-6\mathbf{j}+9\mathbf{k}$$

Therefore

$$|\mathbf{p}_\times\mathbf{q}| = \sqrt{4+36+81}$$

$$= \sqrt{121} = 11$$

Therefore the unit vector is $\frac{1}{11}(2\mathbf{i}-6\mathbf{j}+9\mathbf{k})$

Example 2

The cross product provides us with a spectacular way of solving the area of a triangle or of a parallelogram. Suppose $A(4, 1, -3)$, $B(12, -5, -1)$ and $C(-2, 7, -4)$ are the vertices of a triangle. What is its area?

The area of $ABC = \frac{1}{2}bc \sin A$ which is half the magnitude of a suitable cross product.

$$\mathbf{AB} = [8, -6, 2] \quad \text{and} \quad \mathbf{AC} = [-6, 6, -1]$$

$$\mathbf{AB}_\times\mathbf{AC} = \begin{vmatrix} \mathbf{i} & \mathbf{j} & \mathbf{k} \\ 8 & -6 & 2 \\ -6 & 6 & -1 \end{vmatrix}$$

$$= -6\mathbf{i}-4\mathbf{j}+12\mathbf{k}$$

Therefore area $ABC = \frac{1}{2}|\mathbf{AB}_\times\mathbf{AC}|$

$$= \tfrac{1}{2}\sqrt{36+16+144} = \tfrac{1}{2}\sqrt{196} = 7.$$

We next consider two applications to mechanics of the cross product.

Figure 2.11 shows a force, **F**, acting at a perpendicular distance *ON* from a point *O*.

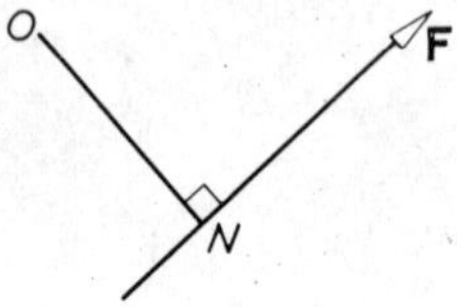

Figure 2.11

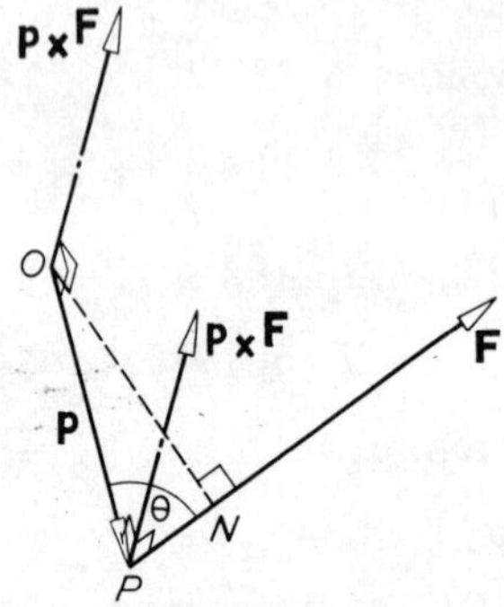

Figure 2.12

In two dimensions, we define the moment of **F** about *O* as the product of **F** and *ON*. In three dimensions, however, we define the moment of **F** about an axis and not about a point. This axis is perpendicular to the plane containing **F** and the vector, **p**, which is the vector relative to *O* of any point, *P*, in the line of action of the force (Figure 2.12).

Therefore **M**, the moment of **F** about an axis through *O*, is defined by

$$\mathbf{M} = \mathbf{p} \times \mathbf{F}$$

The magnitude of this cross product is

$$|\mathbf{p} \times \mathbf{F}| = |\mathbf{F}\mathbf{p}| \sin\theta = |\mathbf{F}|\, ON$$

and the direction of the unit vector is perpendicular to the plane of **p** and **F**.

Example 3

Suppose a force of 30 N acts through the point $P(4, 2, -1)$ in the direction of vector $[2, -5, -14]$. What is the moment of this force about an axis through the point $O(3, 4, 5)$?

Let the force be represented by **F**, then

$$\mathbf{F} = \frac{30}{\sqrt{2^2+(-5)^2+(-14)^2}}(2\mathbf{i}-5\mathbf{j}-14\mathbf{k})$$

$$= \frac{30}{\sqrt{225}}(2\mathbf{i}-5\mathbf{j}-14\mathbf{k}) = 4\mathbf{i}-10\mathbf{j}-28\mathbf{k}$$

$$\mathbf{p} = \mathbf{OP} = (4-3)\mathbf{i}+(2-4)\mathbf{j}+(-1-5)\mathbf{k}$$

$$= \mathbf{i}-2\mathbf{j}-6\mathbf{k}$$

Therefore $$\mathbf{p}_{\times}\mathbf{F} = \begin{vmatrix} \mathbf{i} & \mathbf{j} & \mathbf{k} \\ 1 & -2 & -6 \\ 4 & -10 & -28 \end{vmatrix}$$

$$= 2\{(28-30)\mathbf{i}-(-14+12)\mathbf{j}+(-5+4)\mathbf{k})\}$$

That is $$\mathbf{M} = -4\mathbf{i}+4\mathbf{j}-2\mathbf{k}$$

If there are a number of forces through P, then the moment of their resultant about O is $\mathbf{p}_{\times}\mathbf{R}$ where $\mathbf{R}$ is the sum of the forces. Hence

$$\mathbf{p}_{\times}\mathbf{R} = \mathbf{p}_{\times}(\mathbf{F}_1+\mathbf{F}_2+\mathbf{F}_3+\cdots)$$

or $$\mathbf{p}_{\times}\mathbf{R} = \mathbf{p}_{\times}\mathbf{F}_1+\mathbf{p}_{\times}\mathbf{F}_2+\mathbf{p}_{\times}\mathbf{F}_3+\cdots$$

Figure 2.13

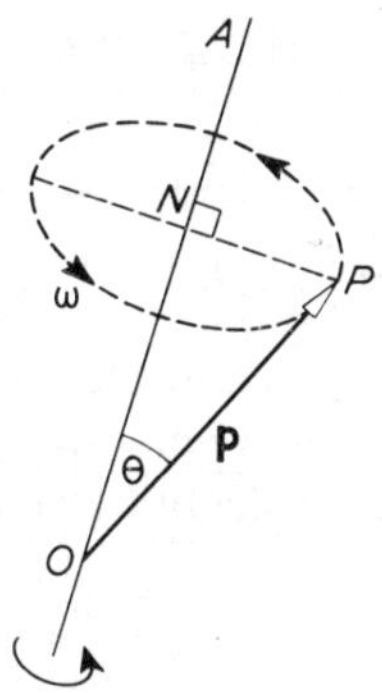

A second application occurs when we consider the angular velocity of a rigid body about a fixed axis. Figure 2.13 shows the axis OA of a body rotating with angular velocity $\boldsymbol{\omega}$.

Let a point P on the body have vector $\mathbf{p}$ relative to O. What is the velocity $\mathbf{v}$ of P?

The magnitude of $\mathbf{v}$ is $\omega . PN$, where N is the foot of the perpendicular drawn from P to the axis OA of the rotating body. That is the magnitude is $\omega|\mathbf{p}|\sin\theta$.

The direction is perpendicular to the plane of OPN and, therefore, if we define the angular velocity of the body as $\boldsymbol{\omega}$, with magnitude ω, in the direction of motion of a right-handed screw, we have

$$\mathbf{v} = \boldsymbol{\omega} \times \mathbf{p}$$

Example 4

Suppose a rigid body is spinning at 22 radians second^{-1} about $O(6, 1, 3)$ parallel to vector $6\mathbf{i}-6\mathbf{j}+7\mathbf{k}$. What is the velocity of a particle at the point $(2, 6, -3)$?

The unit vector in the direction of the angular velocity, $\boldsymbol{\omega}$, is

$$\hat{\boldsymbol{\omega}} = \frac{1}{\sqrt{121}}(6\mathbf{i}-6\mathbf{j}+7\mathbf{k})$$

Therefore $$\boldsymbol{\omega} = \frac{22}{11}(6\mathbf{i}-6\mathbf{j}+7\mathbf{k})$$

$$= 2(6\mathbf{i}-6\mathbf{j}+7\mathbf{k}) \quad \text{or} \quad 12\mathbf{i}-12\mathbf{j}+14\mathbf{k}$$

Hence $$\mathbf{v} = \boldsymbol{\omega} \times \mathbf{p} = 2\begin{vmatrix} \mathbf{i} & \mathbf{j} & \mathbf{k} \\ 6 & -6 & 7 \\ -4 & 5 & -6 \end{vmatrix}$$

$$= 2\{(36-35)\mathbf{i}-(-36+28)\mathbf{j}+(30-24)\mathbf{k}\}$$

That is $$\mathbf{v} = 2\mathbf{i}+16\mathbf{j}+12\mathbf{k}.$$

Such motion as we have seen in these last two examples is neatly described by vectors. Though we have not proved it, the vectors here satisfy the axioms of vector algebra given on page 40.

Summary

The dot product of two vectors $\mathbf{p}$ and $\mathbf{q}$ is defined by

$$\mathbf{p}.\mathbf{q} = p_x q_x + p_y q_y + p_z q_z = pq\cos\theta$$

where θ is the angle between $\mathbf{p}$ and $\mathbf{q}$.

The dot product obeys the following laws

1. $\mathbf{p}.\mathbf{q} = \mathbf{q}.\mathbf{p}$ *Commutative*
2. $\mathbf{p}.(\mathbf{q}+\mathbf{r}) = \mathbf{p}.\mathbf{q}+\mathbf{p}.\mathbf{r}$ *Distributive*
3. $\mathbf{p}.(m\mathbf{q}) = (m\mathbf{p}).\mathbf{q} = m(\mathbf{p}.\mathbf{q})$ *Distributive*

The cross product of two vectors **p** and **q** is defined by

$$\mathbf{p}_\times\mathbf{q} = \begin{vmatrix} \mathbf{i} & \mathbf{j} & \mathbf{k} \\ p_x & p_y & p_z \\ q_x & q_y & q_z \end{vmatrix} \quad \text{symbolically}$$

or $\mathbf{p}_\times\mathbf{q} = pq\sin\theta\,\hat{\mathbf{n}}$

where θ is the angle between **p** and **q**. $\hat{\mathbf{n}}$ is a unit vector in the direction perpendicular to the plane containing **p** and **q** and in the sense of the motion of a right-handed screw turning from **p** through the smaller angle to **q**.

The cross product obeys the following laws

1. $\mathbf{p}_\times\mathbf{q} = -\mathbf{q}_\times\mathbf{p}$ — *Non-commutative*
2. $\mathbf{p}_\times(\mathbf{q}+\mathbf{r}) = \mathbf{p}_\times\mathbf{q}+\mathbf{p}_\times\mathbf{r}$ — *Distributive*
3. $\mathbf{p}_\times(m\mathbf{q}) = (m\mathbf{p})_\times\mathbf{q} = m(\mathbf{p}_\times\mathbf{q})$ — *Distributive*

Exercise 8

1 If $\mathbf{a} = 2\mathbf{i}+4\mathbf{j}+6\mathbf{k}$, $\mathbf{b} = \mathbf{i}+\mathbf{j}$ and $\mathbf{c} = -\mathbf{i}+\mathbf{k}$, find
(i) $\mathbf{a}_\times\mathbf{b}$ (ii) $\mathbf{b}_\times\mathbf{c}$ (iii) $\mathbf{a}_\times\mathbf{c}$

2 Use the vectors of question 1 to verify that
(i) $\mathbf{a}_\times\mathbf{b} = -(\mathbf{b}_\times\mathbf{a})$ (ii) $\mathbf{b}_\times\mathbf{c} = -(\mathbf{c}_\times\mathbf{b})$ (iii) $\mathbf{a}_\times\mathbf{c} = -(\mathbf{c}_\times\mathbf{a})$

3 Use the data of question 1 to verify that
(i) $\mathbf{a}_\times(\mathbf{b}+\mathbf{c}) = \mathbf{a}_\times\mathbf{b}+\mathbf{a}_\times\mathbf{c}$ (ii) $(\mathbf{b}+\mathbf{c})_\times\mathbf{a} = \mathbf{b}_\times\mathbf{a}+\mathbf{c}_\times\mathbf{a}$

4 Show also that $3(\mathbf{a}_\times\mathbf{b}) = (3\mathbf{a})_\times\mathbf{b} = \mathbf{a}_\times(3\mathbf{b})$.

5 Show further that $\mathbf{a}_\times(\mathbf{b}_\times\mathbf{c}) = (\mathbf{a}.\mathbf{c})\mathbf{b}-(\mathbf{a}.\mathbf{b})\mathbf{c}$.

6 The vertices of triangle ABC are $A(-2, 0)$, $B(1, -1)$ and $C(2, 2)$. What are the lengths of the sides of the triangle and its area?

7 Triangle ABC has its vertices $A(2, -1, 3)$, $B(-1, 0, 3)$ and $C(2, 2, 2)$. Find the angles A, B and C, the lengths of the sides and the area of ABC.

8 Find a unit vector perpendicular to each of the vectors $2\mathbf{i}+2\mathbf{j}+\mathbf{k}$ and $2\mathbf{i}+\mathbf{j}$.

9 Find two unit vectors which make an angle of 60° with vector **i** and 45° with vector **k**.

10 Find the area of triangle $A(-1, 0, 1)$, $B(2, -1, 0)$ and $C(0, 1, -1)$. Find also the area of parallelogram A, B, C and $D(-3, 2, 0)$. Is it necessary to know the coordinates of D?

11 Find the sine of the angle between $\mathbf{p} = \mathbf{i}+\mathbf{j}+\mathbf{k}$ and $\mathbf{q} = \mathbf{i}-\mathbf{j}-\mathbf{k}$.

12 Give three possible meanings to the vector equation $\mathbf{a}\times\mathbf{b} = \mathbf{0}$.

13 Prove that $|\mathbf{p}\times\mathbf{q}|^2 = |\mathbf{p}|^2|\mathbf{q}|^2-(\mathbf{p}.\mathbf{q})^2$.

14 Use the data of question 1 to find
(i) $(\mathbf{a}\times\mathbf{b})\times\mathbf{c}$ (ii) $(\mathbf{a}\times\mathbf{b}).(\mathbf{a}\times\mathbf{c})$ (iii) $\{(\mathbf{a}\times\mathbf{b})\times\mathbf{b}\}.\mathbf{c}$

15 A force of 5 newtons acts through $A(3, -4, -5)$ in the direction of vector $2\mathbf{i}-6\mathbf{j}+9\mathbf{k}$. What is the moment of the force about $B(4, 1, 1)$?

16 A rigid body is spinning at 9 radians second^{-1} about an axis through $O(2, -1, 3)$ and which is in the direction of vector $2\mathbf{i}-\mathbf{j}-2\mathbf{k}$. What is the velocity of a particle at the point $P(3, 0, 4)$?
What is the speed of the point and in which direction is its velocity vector?

3 · *Planes, straight lines and space curves*

Planes

What is a plane?

Our definition is based on the property that through a given point P, there exists a unique plane Π perpendicular to a given line, the normal n. All lines through the given point in this plane are orthogonal, that is at right angles, to this normal (Figure 3.1).

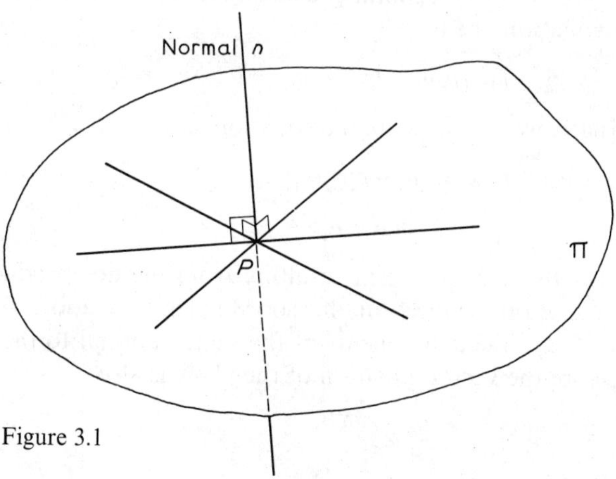

Figure 3.1

Figure 3.2

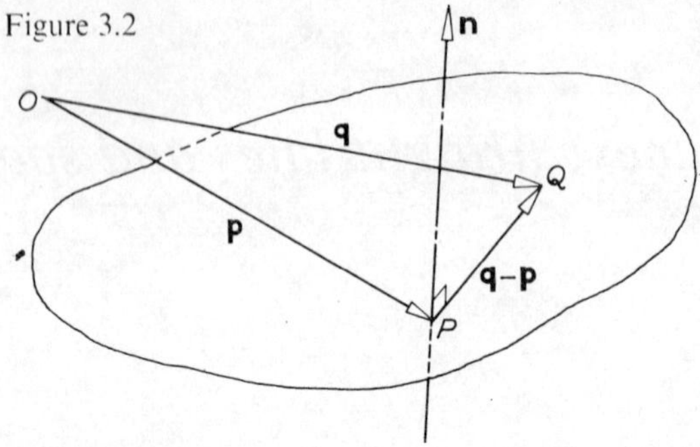

In Figure 3.2, let the given point P have vector **p** relative to an origin O, not necessarily in the plane. The vector perpendicular to the plane is **n**. If Q is any point in the plane with vector **q**, then we know that the dot product **PQ**.**n** is zero, because the angle between the vectors is 90°.

That is $(\mathbf{q}-\mathbf{p}).\mathbf{n} = 0$

This is one form of the equation of a plane since any point on the plane satisfies the equation and no point off the plane satisfies it.

If we give the variable point Q coordinates (x, y, z) and then write out the equation, we have

$$(x-p_x)n_x+(y-p_y)n_y+(z-p_z)n_z = 0 \qquad \ldots \qquad \ldots \text{I}$$

Alternatively we can write the equation as

$$n_x x+n_y y+n_z z-(p_x n_x+p_y n_y+p_z n_z) = 0$$

that is $$n_x x+n_y y+n_z z = c \qquad \ldots \qquad \ldots \text{II}$$

where c is the constant term resulting from the dot product **p**.**n**.

This equation is linear in the coordinates x, y and z and the coefficients are the components of the vector normal to the plane.

Therefore the vector equation of the plane is also

$$\mathbf{q}.\mathbf{n} = c$$

or $\mathbf{q}.\mathbf{n}-c = 0$

where, as before, $\mathbf{p}.\mathbf{n} = c$.

Example 1

Suppose a plane goes through the point (3, −2, 2) and is perpendicular to the vector $6\mathbf{i}-7\mathbf{j}+6\mathbf{k}$. The equation of the plane is

$$6(x-3)-7(y+2)+6(z-2) = 0 \qquad \text{from I}$$

$$6x-18-7y-14+6z-12 = 0$$

that is $$6x-7y+6z = 44$$

Alternatively we can say that the equation of the plane must be of the form

$$6x-7y+6z = c \qquad \text{from II}$$

since the coefficients are components of the normal vector. We know that the point (3, −2, 2) lies on the plane, hence

$$6.3-7(-2)+6.2 = c$$

whence $$44 = c$$

Therefore $6x-7y+6z = 44$

Example 2

Find the equation of the plane through the points $P_1(1, 0, 1)$, $P_2(2, -1, 1)$ and $P_3(0, 1, -1)$.

$$\mathbf{P_1P_2} = [1, -1, 0] \qquad \text{and} \qquad \mathbf{P_3P_2} = [2, -2, 2]$$

We now use that property of the cross product which defines a vector perpendicular to two given vectors. A vector normal to $\mathbf{P_1P_2}$ and $\mathbf{P_3P_2}$ is

$$\begin{vmatrix} x & y & z \\ 1 & -1 & 0 \\ 2 & -2 & 2 \end{vmatrix} = c$$

that is $x(-2)-y(2)+z(-2+2) = c$

$$-2x-2y = c$$

Since the plane goes through the point (1, 0, 1), we have

$$-2(1)-2(0) = c$$

that is $$c = -2$$

whence we have the equation of the plane

$$x+y = 1$$

Perpendicular distance from a point to a plane

The perpendicular distance from a point to a plane is parallel to the normal vector **n** and the simplest case occurs when the point is at the origin, O.

Figure 3.3

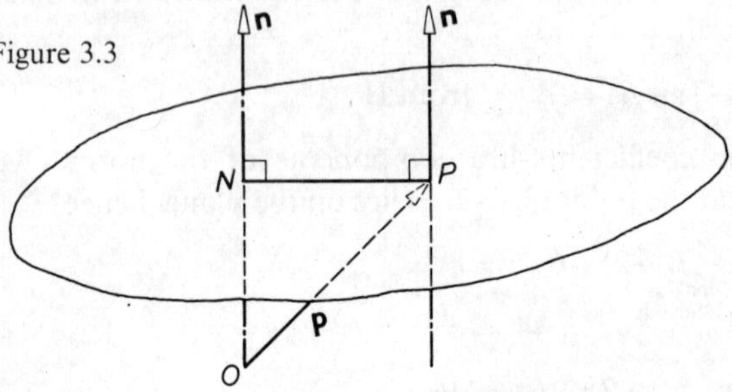

We require the distance ON (Figure 3.3). How do we calculate ON?

We have already nearly done so because the dot product $\mathbf{p}.\mathbf{n}$ is the projection of **p** on to **n**. All we have to do, therefore, is to divide $\mathbf{p}.\mathbf{n}$ by the magnitude of **n**, that is n.

$\mathbf{n}/n$ is the unit normal vector, $\hat{\mathbf{n}}$, to the plane. We have, therefore, the distance from the origin ON, where

$$ON = \left|\frac{\mathbf{p}.\mathbf{n}}{n}\right| = \left|\mathbf{p}.\hat{\mathbf{n}}\right|$$

Figure 3.4

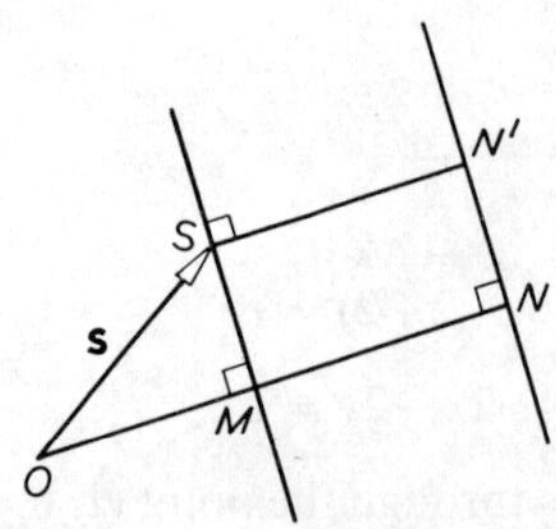

If, on the other hand, we have another point S (Figure 3.4), vector **s** relative to O, then the distance we now require is SN'. SM is the plane through S parallel to the plane $N'N$. Here we regard the planes as perpendicular to the plane of the paper so that their lines of intersection with the paper are SM and $N'N$.

$$SN' = ON - OM$$

$$= \left|\frac{\mathbf{p}.\mathbf{n}}{n} - \frac{\mathbf{s}.\mathbf{n}}{n}\right|$$

$$= |\mathbf{p}.\hat{\mathbf{n}} - \mathbf{s}.\hat{\mathbf{n}}|$$

$$= |(\mathbf{p} - \mathbf{s}).\hat{\mathbf{n}}|$$

Example

What is the distance from (1) the origin and (2) the point $(2, -3, 3)$ to the plane through $(-3, 4, -1)$ and parallel to the plane $6x - 2y + 3z = 5$?

(1) $\mathbf{p} = [-3, 4, -1]$

$$\hat{\mathbf{n}} = \frac{[6, -2, 3]}{\sqrt{6^2 + (-2)^2 + 3^2}} = \tfrac{1}{7}[6, -2, 3]$$

Therefore the distance d_1 of the origin from the plane is

$$d_1 = |\mathbf{p}.\hat{\mathbf{n}}| = |[-3, 4, -1].\tfrac{1}{7}[6, -2, 3]|$$

$$= \tfrac{1}{7}|(-18 - 8 - 3)| = 4\tfrac{1}{7}.$$

(2) $\mathbf{p} - \mathbf{s} = [\{2 - (-3)\}, (-3 - 4), \{3 - (-1)\}]$

$$= [5, -7, 4]$$

$$\hat{\mathbf{n}} = \tfrac{1}{7}[6, -2, 3]$$

Therefore the distance d_2 of the point $(2, -3, 3)$ from the plane is

$$d_2 = |(\mathbf{p} - \mathbf{s}).\mathbf{n}| = \tfrac{1}{7}[5, -7, 4].[6, -2, 3]$$

$$= \tfrac{1}{7}(30 + 14 + 12) = 8.$$

The angle between two planes

Let the two planes be perpendicular to the plane of the paper and their intersections with the paper be APA' and BPB'. If their respective orthogonal vectors are $\mathbf{PN}_1$ and $\mathbf{PN}_2$, then these are in the plane of the paper (Figure 3.5).

Figure 3.5

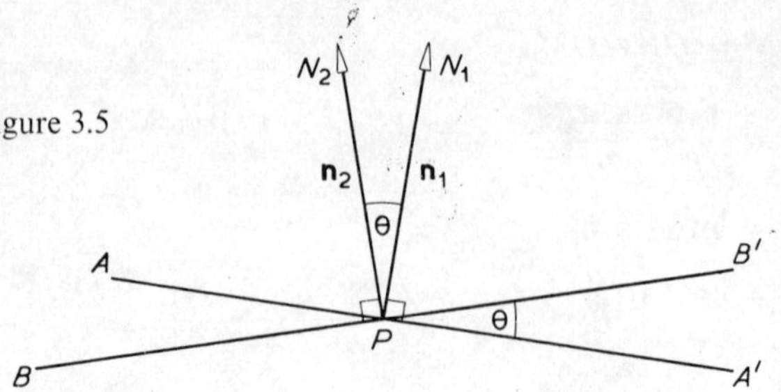

Further $\angle A'PB' = \angle A'PN_1 - \angle B'PN_1 = 90° - \angle B'PN_1$
$= \angle B'PN_2 - \angle B'PN_1 = \angle N_1PN_2$
that is the angle between the planes equals the angle between their normals.

If, therefore, their respective normals are $\mathbf{n}_1$ and $\mathbf{n}_2$, we can obtain the angle between them θ by calculating their scalar product

$$\mathbf{n}_1 \cdot \mathbf{n}_2 = |\mathbf{n}_1||\mathbf{n}_2| \cos \theta$$

whence $\cos \theta = \dfrac{\mathbf{n}_1 \cdot \mathbf{n}_2}{n_1 \cdot n_2} = \hat{\mathbf{n}}_1 \cdot \hat{\mathbf{n}}_2$

Example

What is the angle between the two planes $6x - 7y + 6z = 44$ and $6x - 2y + 3z = 21$?

$$\mathbf{n}_1 \cdot \mathbf{n}_2 = 6(6) + (-7)(-2) + 6(3) = 68$$
$$n_1 n_2 = \sqrt{36+49+36}\sqrt{36+4+9} = (11)(7) = 77$$

Therefore $\cos \theta = \dfrac{68}{77}$ and $\theta = 27° \, 59'$.

Exercise 9

1 If $\mathbf{a} = 3\mathbf{i} + 2\mathbf{j} + \mathbf{k}$ and $\mathbf{b} = \mathbf{i} - \mathbf{j} + \mathbf{k}$, show that $\mathbf{a} - \mathbf{b}$ is parallel to the x, y plane.

2 If $\mathbf{p} = \mathbf{i} + \mathbf{j} + \mathbf{k}$ and $\mathbf{q} = \mathbf{i} - \mathbf{j} - \mathbf{k}$, to what plane is $\mathbf{p} - \mathbf{q}$ parallel?

3 Find the equations of the planes containing the following given points and normal to the given directions
(i) $A(a_1, a_2, a_3)$ and normal to the unit vector $\hat{\mathbf{n}}$
(ii) $A(2, 1, -3)$ and normal to vector $\mathbf{i} - 2\mathbf{j} - 3\mathbf{k}$
(iii) $A(0, 4, 3)$ and normal to vector $2\mathbf{i} + 2\mathbf{j} + \mathbf{k}$

4 Find the angle between the following pairs of planes
(i) $x+2y+2z = 0$ and $x+z = 0$
(ii) $x+y+2z = 1$ and $x+2y+z = 0$

5 Find the equation of the plane through the point $(2, 3, -4)$ and parallel to the plane $3x-4y-3z = 5$.

6 Show that the distance of the point $P(x, y)$ from the line $ax+by+c = 0$ is given by $d = |\hat{\mathbf{n}}.\mathbf{QP}|$ where $\hat{\mathbf{n}}$ is the unit normal to $ax+by+c = 0$ and Q is a point on the line.

7 Calculate the distance of $P(5, 4)$ from the line $3x-4y+6 = 0$.

8 (i) What is the distance of the point $(1, 1, 2)$ from the plane $x+2y+2z = 3$?
(ii) What is the distance of the point $(1, -2, -1)$ from the plane $x+y+3z+2 = 0$?

9 Find the equations of the planes determined by the points
(i) $P_1(0, 0, 1)$, $P_2(1, -1, 0)$ and $P_3(2, 1, -1)$
(ii) $P_1(2, 0, -3)$, $P_2(3, -2, -2)$ and $P_3(5, -3, -1)$

10 Find the Cartesian equation of the sphere, centre $(3, 2, 5)$, which has the plane $2x+2y+z = 6$ tangential to it.

The straight line

What determines that a set of points shall lie on a straight line? We discuss three important conditions.

1. *A point and vector*

Figure 3.6

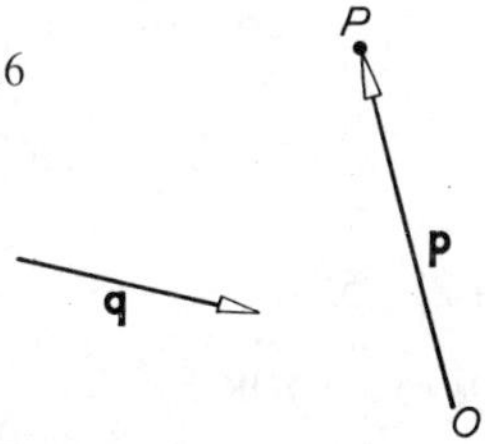

Figure 3.6 shows a point P, vector $\mathbf{p}$ and vector $\mathbf{q}$, both relative to the origin O.

Suppose we draw a line through P parallel to $\mathbf{q}$ (Figure 3.7). The line is unlimited in length. How can we arrive at any point R on this line?

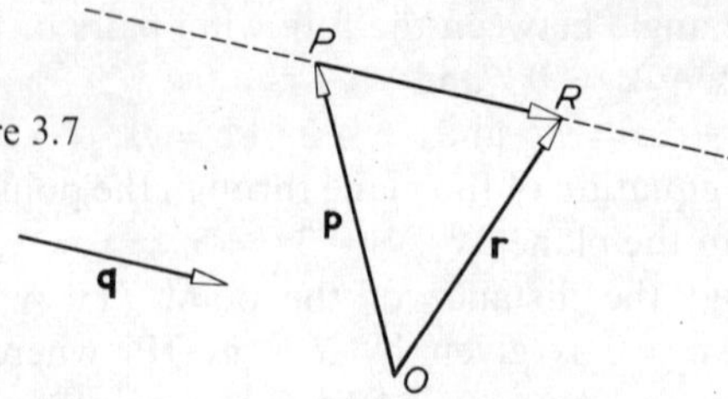

Figure 3.7

From Figure 3.7

$$\mathbf{r} = \mathbf{p} + \mathbf{PR}$$

Though **PR** has the same direction as **q**, its length is determined by wherever the point R is situated. The equation, therefore, must be sufficiently flexible to satisfy this condition. In such cases, we employ a *parameter* to accommodate R and write

$$\mathbf{PR} = t\mathbf{q}$$

Here t is the parameter: it is a real number employed to describe the distance PR. Parameter t is constant for a given length but varies as the distance PR changes to enable us to consider different lengths.

For the first equation of a straight line, therefore, we have

$$\mathbf{r} = \mathbf{p} + t\mathbf{q}$$

This equation represents the straight line through a given point P and parallel to a given vector **q**.

Example 1

Let the given point be $P(1, -2, 3)$ and the given vector be $\mathbf{q} = 4\mathbf{i} + 2\mathbf{j} - 5\mathbf{k}$.

$$\begin{aligned} \mathbf{r} &= \mathbf{p} + t\mathbf{q} \\ &= \mathbf{i} - 2\mathbf{j} + 3\mathbf{k} + t(4\mathbf{i} + 2\mathbf{j} - 5\mathbf{k}) \\ &= (1+4t)\mathbf{i} - (2-2t)\mathbf{j} + (3-5t)\mathbf{k} \end{aligned}$$

If we use the alternative notation, we have

$$\mathbf{r} = [1+4t,\ -2+2t,\ 3-5t]$$

In coordinate form, this is

$$x = 1+4t, \qquad y = -2+2t, \qquad z = 3-5t$$

which may be written as

$$\frac{x-1}{4} = \frac{y+2}{2} = -\frac{z-3}{5} = t,$$

the standard form of the equations of a straight line.

How can we determine whether a point is on the line or not?

Suppose the point is (9, 2, −7). The coordinates of this point must satisfy the corresponding coordinate components we have already found for the straight line vector equation, that is

$$\begin{aligned} x &= 1+4t = 9, \quad \text{whence} \quad t = 2 \\ y &= -2+2t = 2, \quad \text{whence} \quad t = 2 \\ z &= 3-5t = -7, \quad \text{whence} \quad t = 2 \end{aligned}$$

Hence the parameter t has the same value at this point. Therefore the equation is satisfied and the point is on the line.

Alternatively once we have a value for t from the parametric equation involving x in terms of t, we can substitute this value in the equations for y and z to determine whether their respective values agree with the y and z values of the point.

For example when $t = 2$,

$$y = -2+2(2) = 2$$

which agrees with the y value of the point.

Again when $t = 2$,

$$z = 3-5(2) = -7$$

which agrees with the z value of the point.

Example 2

Where does the line

$$\frac{x-1}{4} = \frac{y+2}{2} = -\frac{z-3}{5} = t$$

intersect a plane, for instance the plane $x = 5$?

When $x = 5$, we know that from the equation of the line,

$$x = 1+4t$$

Therefore $5 = 1+4t$ and $t = 1$

When we substitute this value of t in the equations of the line, we have

$$y = -2+2(1) = 0$$

$$z = 3-5(1) = -2$$

Therefore the coordinates are $(5, 0, -2)$.

On the other hand if we are given the equation of a plane in the form

$$x+2y+z+9 = 0$$

we then substitute the parameter for the coordinates

$$(1+4t)+2(-2+2t)+(3-5t)+9 = 0$$

whence $t = -3$

Therefore $x = 1+4(-3) = -11$

$$y = -2+2(-3) = -8$$

$$z = 3-5(-3) = 18$$

and the coordinates are $(-11, -8, 18)$.

2. *Two given points*

Figure 3.8

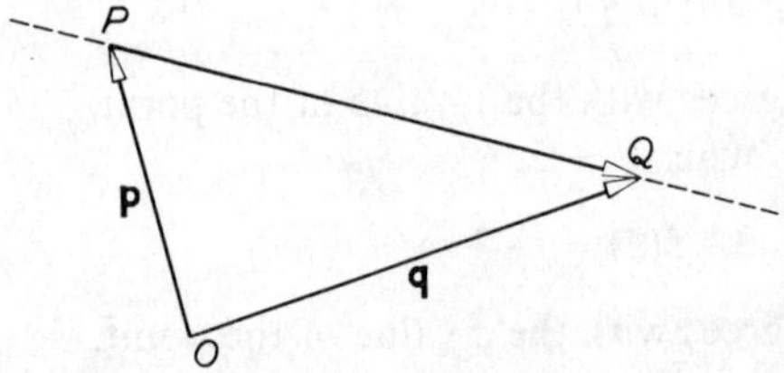

Figure 3.8 shows two points P and Q with vectors $\mathbf{p}$ and $\mathbf{q}$ relative to O.

The line joining these two points has the direction of vector $\mathbf{PQ}$, $\mathbf{q}-\mathbf{p}$. Again, however, the length of the line is not limited. Therefore a parametric equation for the straight line through P and Q is

$$\mathbf{r} = \mathbf{p}+t(\mathbf{q}-\mathbf{p})$$

or $\mathbf{r} = (1-t)\mathbf{p}+t\mathbf{q}$

We recall from page 26 that if we write this equation in the form

$$(1-t)\mathbf{p}+t\mathbf{q}-\mathbf{r} = 0$$

a condition for collinearity of points is that the coefficients shall be zero. In this case

$$1-t+t-1 = 0$$

so that we have confirmation that the point Q is collinear with P.

Example

Suppose two points are $P(1, 0, 1)$ and $Q(3, 2, 1)$.

$$\mathbf{PQ} = \mathbf{q}-\mathbf{p} = (3-1)\mathbf{i}+(2-0)\mathbf{j}+(1-1)\mathbf{k}$$

$$= 2\mathbf{i}+2\mathbf{j}$$

Therefore $\mathbf{r} = \mathbf{p}+t(\mathbf{q}-\mathbf{p}) = \mathbf{i}+\mathbf{k}+t(2\mathbf{i}+2\mathbf{j})$

that is $\quad \mathbf{r} = (1+2t)\mathbf{i}+2t\mathbf{j}+\mathbf{k}$

Alternative ways of writing this equation are

$$\mathbf{r} = [1, 0, 1]+t[2, 2, 0]$$

or $\quad \mathbf{r} = [1+2t, 2t, 1]$

3. *Two non-parallel planes*

When two planes intersect, what is the figure formed along the points of intersection?

A little thought soon convinces us that it is a straight line. A third way, therefore, to find an equation of a straight line is to use the property that two intersecting planes determine a straight line.

Figure 3.9

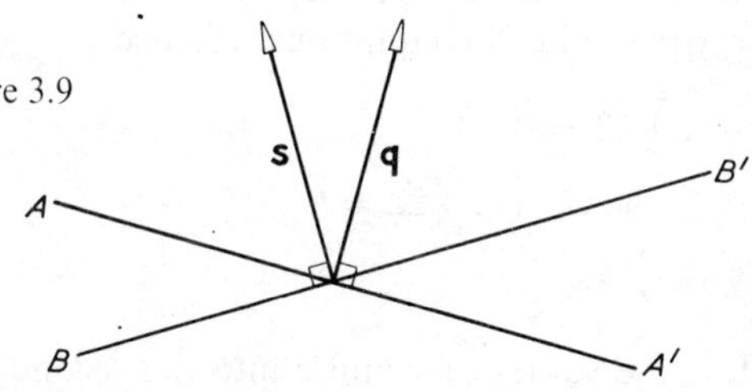

Suppose two planes, APA' and BPB', have orthogonal vectors $\mathbf{q}$ and $\mathbf{s}$ (Figure 3.9). We assume as we have done on other occasions that this diagram represents two planes perpendicular to the plane of the paper and intersecting the paper in AA' and BB'. Therefore the orthogonal vectors $\mathbf{q}$ and $\mathbf{s}$ are in the plane of the paper.

The cross product $\mathbf{q}_\times\mathbf{s}$ defines a vector perpendicular to both these vectors and, therefore, perpendicular to the plane of the paper: that is, the cross product defines a vector coincident with the intersection of the planes.

Hence for a given point P, with vector $\mathbf{p}$, we have the third equation

$$\mathbf{r} = \mathbf{p} + t(\mathbf{q}_\times\mathbf{s})$$

Example

Let two intersecting planes have equations

$$2x - y + 3z + 8 = 0$$

$$x + 5y - z - 2 = 0$$

From the first equation the orthogonal vector $\mathbf{q}$, say, is $2\mathbf{i} - \mathbf{j} + 3\mathbf{k}$, and from the second equation the orthogonal vector $\mathbf{s}$ is $\mathbf{i} + 5\mathbf{j} - \mathbf{k}$.

$$\mathbf{q}_\times\mathbf{s} = \begin{vmatrix} \mathbf{i} & \mathbf{j} & \mathbf{k} \\ 2 & -1 & 3 \\ 1 & 5 & -1 \end{vmatrix}$$

$$= -14\mathbf{i} + 5\mathbf{j} + 11\mathbf{k}$$

We now require a point P common to both planes. There are an infinite number of points but we need only one.

Eliminate x from the two equations, whence

$$-11y + 5z + 12 = 0$$

If we set $z = 2$, then $-11y + 22 = 0$

and $\quad y = 2$

Substitute these values of y and z into the second equation and we have

$$x + 10 - 2 - 2 = 0$$

and $\quad x = -6$

Therefore, one point common to both planes is $P(-6, 2, 2)$

We now have an equation to represent the straight line which is the intersection of the two planes

$$\mathbf{r} = [-6, 2, 2]+t[-14, 5, 11]$$

Alternatively, let the direction cosines of the line be l, m and n. The normals of the two planes have direction ratios

$$2:-1:3 \quad \text{and} \quad 1:5:-1$$

The line of intersection is perpendicular to each of these directions.

Therefore $\quad 2l-m+3n = 0$

and $\quad l+5m-n = 0$

whence $\quad l:m:n = -14:5:11$

and $\quad l = \dfrac{-14}{\sqrt{342}}, \quad m = \dfrac{5}{\sqrt{342}}, \quad n = \dfrac{11}{\sqrt{342}}$

We know that a point on the line is $P(-6, 2, 2)$.

Therefore the standard form of the equation of the straight line is

$$\frac{x+6}{-14} = \frac{y-2}{5} = \frac{z-2}{11} = t = \frac{r}{\sqrt{342}}$$

where r is the distance of the point (x, y, z) from the point $P(-6, 2, 2)$.

Again we can use instead two points from the intersection of the two planes. Having eliminated x from the two equations, we then set $z = 13$, whence

$$-11y+65+12 = 0$$

and $\quad y = 7$

Substitute these values of y and z into the second equation of the planes, and we have

$$x+35-13-2 = 0$$

and $\quad x = -20$

Therefore a second point common to both planes is $Q(-20, 7, 13)$. The equation of the straight line is now in the form

$$\mathbf{r} = \mathbf{p}+t(\mathbf{q}-\mathbf{p})$$

and we have

$$\mathbf{r} = [-6, 2, 2]+t[-14, 5, 11]$$

Exercise 10

1 Find a parametric equation of the line through the point (1, 2, 3) and parallel to vector $2\mathbf{i}-4\mathbf{j}+7\mathbf{k} = 5$.

2 Find an equation of the straight line through the point (3, −2, 5) and perpendicular to the plane $4x+6y-7z = 1$.

3 Give parametric equations of the lines joining the following pairs of points

(i) (a_1, a_2) and (b_1, b_2) (ii) (2, 5) and (5, −1)

(iii) (−4, 0) and (3, 1) (iv) (7, −2) and (−6, 4)

4 Give parametric equations of the lines joining the following pairs of points

(i) (a_1, a_2, a_3) and (b_1, b_2, b_3) (ii) (3, 2, 1) and (−1, 2, 1)

(iii) (1, 0, −2) and (3, 4, −1) (iv) (0, 2, 0) and (1, −1, 2)

5 Find parametric equations of the lines of intersection of the following pairs of planes

(i) $a_1x+a_2y+a_3z+a_4 = 0$ and $b_1x+b_2y+b_3z+b_4 = 0$

(ii) $2x-3y+z-1 = 0$ and $2x+3y-z+2 = 0$

(iii) $3x+y-z = 0$ and $x-y+2z+1 = 0$

(iv) $x+y = 0$ and $y-z = 0$

6 What are the coordinates of the points where the lines of question 5 parts (ii), (iii) and (iv) intersect the three coordinate planes $x = 0$, $y = 0$ and $z = 0$?

7 Find the cosine of the angle between the lines

$x+1 = y = z-1$ and $x = 1-t,\ y = t,\ z = 5t$

8 What is the line of intersection of the planes

$2x-y+z = 5$ and $2x+2y-z = 2$?

9 Find the distances of the points (1, −1, −2) and (3, −2, 6) from the plane which passes through the point (3, −2, 1) and is perpendicular to the line of intersection of the planes $3x+2y-z = 2$ and $2x-y+3z = 3$.

10 Find the line of intersection of the planes

$x-y+2z-6 = 0$ and $2x+y+z-4 = 0$

Show that the plane $x-4y+5z-14 = 0$ intersects the given planes in the same straight line.

Coplanar curves

The Circle

In Cartesian form the equation of a circle, radius a centred on the origin is

$$x^2 + y^2 = a^2$$

Figure 3.10

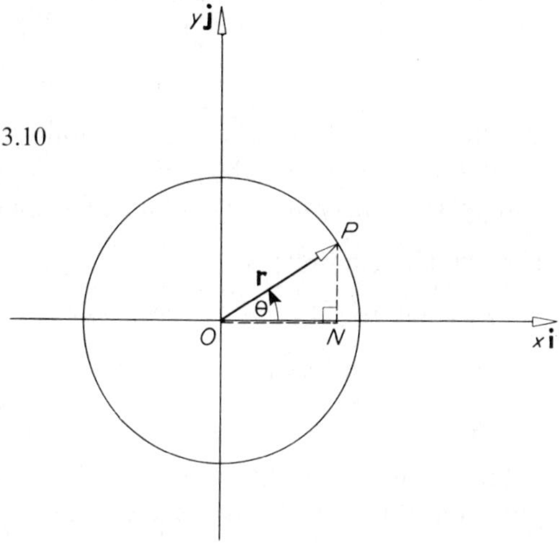

How can we write this in vectors?

If the vector of a point on the circumference is $\mathbf{r}$ (Figure 3.10), then

$$\mathbf{r} = \mathbf{ON} + \mathbf{NP} = x\mathbf{i} + y\mathbf{j}$$
$$= a\cos\theta\,\mathbf{i} + a\sin\theta\,\mathbf{j}$$

where a is the length of the radius and θ is the angle between the radius vector and the x axis.

There are many vector equations of the circle and in this one the parameter θ is given in radians.

A second form is

$$\mathbf{r} = a\cos\left(\frac{s}{a}\right)\mathbf{i} + a\sin\left(\frac{s}{a}\right)\mathbf{j}$$

where s takes the range of values $0 \leqslant s \leqslant 2\pi a$ and is the arc length from $(a, 0)$ to $P(x, y)$ on the circle measured in an anti-clockwise direction.

A third form is

$$\mathbf{r} = a\cos\omega t\,\mathbf{i}+a\sin\omega t\,\mathbf{j}$$

where t is the parameter measured in time for values of $t \geqslant 0$ and ω is the angular velocity of the radius vector.

Though angular velocity is technically a vector quantity, $\boldsymbol{\omega}$, the same phrase is usually used for the scalar speed ω.

We assume that when $t = 0$, the radius vector is at $(a, 0)$.

Example

A point revolves about the origin at 5 revolutions second^{-1} and at a distance of 2 units. Where is the point when $t = \frac{5}{8}$ second?

Since the angular velocity $\omega = 5(2\pi)$ radians second^{-1}, we have

$$\mathbf{r} = 2\cos 10\pi t\,\mathbf{i}+2\sin 10\pi t\,\mathbf{j}$$

When $t = \frac{5}{8}$ second, then

$$\begin{aligned}\mathbf{r} &= 2\cos 10\pi\tfrac{5}{8}\,\mathbf{i}+2\sin 10\pi\tfrac{5}{8}\,\mathbf{j}\\ &= 2\cos 6\tfrac{1}{4}\pi\,\mathbf{i}+2\sin 6\tfrac{1}{4}\pi\,\mathbf{j}\end{aligned}$$

Therefore the radius vector $\mathbf{r}$ has completed $6\frac{1}{4}\pi/2\pi$ or $3\frac{1}{8}$ revolutions. The position of $\mathbf{r}$ in the fourth revolution is

$$\begin{aligned}\mathbf{r} &= 2\cos\tfrac{1}{4}\pi\,\mathbf{i}+2\sin\tfrac{1}{4}\pi\,\mathbf{j}\\ &= \frac{2}{\sqrt{2}}\mathbf{i}+\frac{2}{\sqrt{2}}\mathbf{j} = \sqrt{2}\mathbf{i}+\sqrt{2}\mathbf{j}\end{aligned}$$

An important application of vector equations in parameter t is to velocities and accelerations as well as to positions because, as we shall see in the next chapter, we can find the rates of change of motion by vector differentiation and integration.

Exercise 11

1 What is the vector equation of the locus of all points P such that they lie on a circle with diameter AB?

2 What is the vector equation of the circle with: (i) centre at the origin and radius 2? (ii) centre at (2, 3) and radius 4?

3 Sketch the curves whose vector equations are

(i) $\mathbf{r} = \cos\theta\,\mathbf{i}+\sin\theta\,\mathbf{j}$

(ii) $\mathbf{r} = 3\cos\theta\,\mathbf{i}+3\sin\theta\,\mathbf{j}$

4 Find the centres and lengths of diameters of the circles whose vector equations are

(i) $\mathbf{r} = (1+\cos\theta)\mathbf{i}+(2+\sin\theta)\mathbf{j}$

(ii) $\mathbf{r} = (2\cos\theta-3)\mathbf{i}+2\sin\theta\,\mathbf{j}$

(iii) $\mathbf{r} = (3\cos\theta+1)\mathbf{i}+3\sin\theta\,\mathbf{j}$

5 Obtain the corresponding Cartesian coordinate equations of question 4.

6 For the vector equation $\mathbf{r} = a\cos(s/a)\,\mathbf{i}+a\sin(s/a)\,\mathbf{j}$, find the values of $\mathbf{r}$ for the following values of s

(i) $s = 0$ (ii) $s = \pi a/4$ (iii) $s = \pi a/2$

(iv) $s = 2\pi a/3$ (v) $s = 5\pi a/3$

7 A particle moves anti-clockwise on a circular path, centre at the origin and it begins at the point (2, 0). If the particle completes one revolution every 3 seconds, write a position vector equation of the motion of the particle with respect to time in seconds. Where is the particle after (i) $\frac{1}{2}$ second and (ii) $1\frac{1}{4}$ seconds?

8 Find values of t for which the line $\mathbf{r}_1 = 2\mathbf{i}+t(-\mathbf{i}+\mathbf{j})$ intersects the circle $\mathbf{r}_2 = 2\cos p\,\mathbf{i}+2\sin p\,\mathbf{j}$.
What is the length of the intercept?

9 Find the value of p when the line $\mathbf{r}_1 = 2\mathbf{i}+t(-12\mathbf{i}+5\mathbf{j})$ intersects the circle $\mathbf{r}_2 = \cos p\,\mathbf{i}+\sin p\,\mathbf{j}$.

10 Find values of p for which the line $\mathbf{r}_1 = 2\mathbf{i}+t(-\mathbf{i}+p\mathbf{j})$ is a tangent to the circle $\mathbf{r}_2 = \cos p\,\mathbf{i}+\sin p\,\mathbf{j}$.

11 What can you say about the dot product of $\mathbf{r}_1$ and $\mathbf{r}_2$ for the values of p in question 10?
Illustrate this by finding the value of t for which the line

$$\mathbf{r}_1 = \sqrt{10}\mathbf{i}+t\left(-\mathbf{i}+\frac{\theta}{\pi}\mathbf{j}\right)$$

is a tangent to $\mathbf{r}_2 = \cos\theta\,\mathbf{i}+\sin\theta\,\mathbf{j}$.

The simplest parametric form we can use to satisfy the equation of a parabola

$$y^2 = 4ax$$

is $\quad x = ap^2 \quad$ and $\quad y = 2ap.$

The focus of the parabola is at $(a, 0)$ and the equation of the directrix is $x = -a$.

Figure 3.11

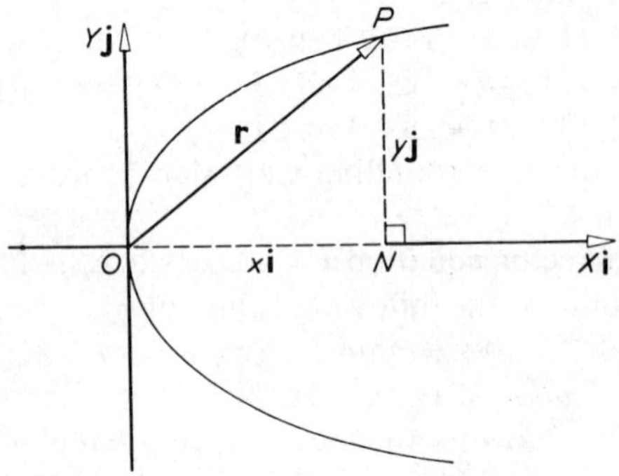

From Figure 3.11 and using the parameter p, therefore, we have the following vector equation for a parabola

$$\begin{aligned} \mathbf{r} &= \mathbf{ON}+\mathbf{NP} = x\mathbf{i}+y\mathbf{j} \\ &= ap^2\mathbf{i}+2ap\mathbf{j} \end{aligned}$$

If we use parameter t as a measure of time, then

$$\mathbf{r} = at^2\mathbf{i}+2at\mathbf{j}$$

Exercise 12

1 Plot the curves whose vector equations are
(i) $\mathbf{r} = p^2\mathbf{i}+p\mathbf{j}$ (ii) $\mathbf{r} = p\mathbf{i}+p^2\mathbf{j}$ (iii) $\mathbf{r} = p^2\mathbf{i}+2p\mathbf{j}$

2 Obtain the corresponding Cartesian coordinate equations of question 1 by the elimination of the parameter p.

3 Draw a graph of $\mathbf{r} = p^2\mathbf{i}+2p\mathbf{j}$
On the graph draw the vector **OA** where O is the origin and A the point corresponding to $p = 2$. Let this vector be $\mathbf{r}(2)$.
Draw now the vector **OB**, corresponding to $\mathbf{r}(3)$.
Draw and state the position of the vector **AB**.

4 Repeat question 3 for the points A and C corresponding to $p = 2$ and $p = 2.1$ respectively.
Draw the vector $\dfrac{\mathbf{r}(2.1)-\mathbf{r}(2)}{0.1}$ and state its position.
Deduce the vector equation of the tangent at $p = 2$.

5 The path of a projectile, if we neglect air resistance, is given approximately by
$$\mathbf{r} = u(\cos\alpha)t\,\mathbf{i}+(u\sin\alpha\,t-\tfrac{1}{2}gt^2)\mathbf{j}$$
where u is the initial velocity, α is the angle of projection, t the time measured from the time of projection and g the acceleration due to gravity which is constant and approximately $9{\cdot}81\ \text{m s}^{-2}$. If $u = 49{\cdot}05\sqrt{2}\ \text{m s}^{-1}$ and $\alpha = 45°$, find
(i) the time when the particle reaches the ground again
(ii) the range of the particle
(iii) the maximum height of the particle
(iv) the path of the projectile drawn on a graph.

6 Find the points of intersection of the line $\mathbf{r}_1 = 3\mathbf{i}+t(\mathbf{i}+\mathbf{j})$ and the curve $\mathbf{r}_2 = p^2\mathbf{i}+2p\mathbf{j}$.

7 Find the values of ϕ for which the line $\mathbf{r}_1 = t(\phi\mathbf{i}+\mathbf{j})-3\mathbf{i}$ intersects the parabola $\mathbf{r}_2 = p^2\mathbf{i}+2p\mathbf{j}$.

8 What are the vectors of the points of intersection of the circle $\mathbf{r}_1 = \cos p\,\mathbf{i}+\sin p\,\mathbf{j}$ with the parabola $\mathbf{r}_2 = p^2\mathbf{i}+2p\mathbf{j}$? What is the length of the straight line segment joining the points of intersection?

9 Find the value of a for which the line $\mathbf{r}_1 = a\mathbf{i}+t(\mathbf{i}+\mathbf{j})$ is a tangent to the parabola $\mathbf{r}_2 = p^2\mathbf{i}+2p\mathbf{j}$. When a has this value, find the distance from the origin to where the tangent touches the parabola.

10 Find the length of the intercept when the straight line $\mathbf{r}_1 = (t+3)\mathbf{i}+t\mathbf{j}$ cuts the curve $\mathbf{r}_2 = p^2\mathbf{i}+2p\mathbf{j}$.

The Ellipse

The standard form of an ellipse is

$$\frac{x^2}{a^2}+\frac{y^2}{b^2} = 1$$

where the major axis is $2a$, minor axis $2b$, focus at $(ae, 0)$, directrix $x = a/e$ and the eccentricity, $e < 1$, found from the relationship $b^2 = a^2(1-e^2)$. The semi-latus rectum is b^2/a.

If we now use the parametric equations

$$x = a\cos\phi \quad \text{and} \quad y = b\sin\phi$$

we can form a vector equation for an ellipse (Figure 3.12)

Figure 3.12

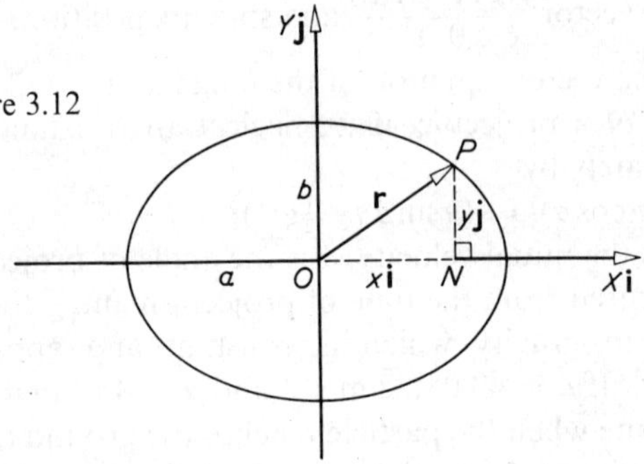

$$\mathbf{r} = \mathbf{ON}+\mathbf{NP} = x\mathbf{i}+y\mathbf{j}$$

$$= a\cos\phi\mathbf{i}+b\sin\phi\mathbf{j}$$

The Hyperbola

Figure 3.13 shows a hyperbola.

Characteristics of the hyperbola are: if $AA' = 2a$, then the focus S is $(ae, 0)$, directrix MM' is $x = a/e$ and the eccentricity can be found from the relationship $b^2 = a^2(e^2-1)$, since for the hyperbola $e > 1$. The semi-latus rectum is b^2/a.

The equation for the hyperbola is

$$\frac{x^2}{a^2}-\frac{y^2}{b^2} = 1$$

When we use the parametric equations

$$x = a\sec\phi \quad \text{and} \quad y = b\tan\phi$$

Figure 3.13

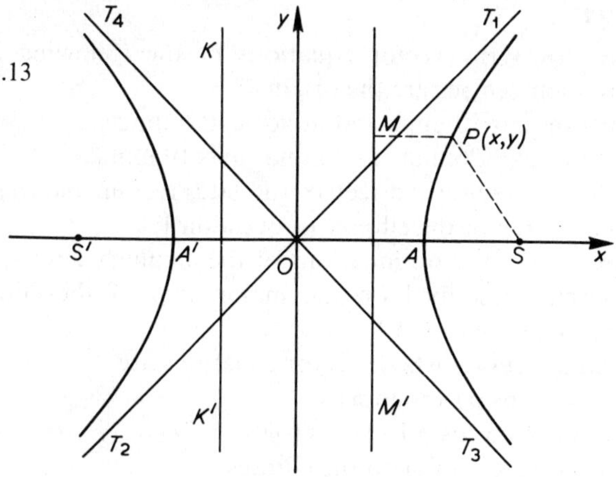

we can then form a vector equation for the hyperbola:

$$\mathbf{r} = a\sec\phi\,\mathbf{i}+b\tan\phi\,\mathbf{j}$$

The asymptotes T_1T_2 and T_3T_4, the lines which meet the hyperbola only at an infinite distance, have equations

$$y = \pm\frac{b}{a}x$$

or $\mathbf{r}_1 = \lambda(a\mathbf{i}+b\mathbf{j})$ and $\mathbf{r}_2 = \mu(a\mathbf{i}-b\mathbf{j})$

When the hyperbola is such that $a = b$, the asymptotes are inclined at angles of $\pm 45°$ to the x axis. Therefore the asymptotes are at right angles to each other, and the hyperbola is called a rectangular hyperbola. The equation of the curve is then more conveniently referred to its asymptotes

$$xy = c^2$$

where $4c^2$ equals the sum of the squares of the semi-axes of the hyperbola.

The parametric equations

$$x = ct \quad \text{and} \quad y = \frac{c}{t}$$

satisfy this equation so that we have

$$\mathbf{r} = ct\mathbf{i}+\frac{c}{t}\mathbf{j} \quad \text{where } t \text{ is a parameter.}$$

Exercise 13

1 Write down the vector equations of the following ellipses where their centres are the origin
(*a*) major axis 10 units and minor axis 6 units
(*b*) major axis 26 units and minor axis 10 units

2 Find the (i) focus, (ii) directrix, (iii) latus rectum and (iv) eccentricity of each of the ellipses of question 1.

3 What are the vector equations of the auxiliary circles, that is the circles described on the major axes of the ellipses as diameters, of question 1?

4 Sketch the curves whose vector equations are
(i) $\mathbf{r}_1 = 3\cos\phi\,\mathbf{i}+4\sin\phi\,\mathbf{j}$
(ii) $\mathbf{r}_2 = (2+3\cos\phi)\mathbf{i}+(1+\sin\phi)\mathbf{j}$

5 For what values of ϕ do the ellipses

$$\mathbf{r}_1 = 5\cos\phi\,\mathbf{i}+3\sin\phi\,\mathbf{j}$$
$$\mathbf{r}_2 = 5\cos 2\phi\,\mathbf{i}+3\sin 2\phi\,\mathbf{j}$$

coincide?

6 Find the length of the straight line intercept made by $\mathbf{r}_1 = t(\mathbf{i}+\mathbf{j})$ with $\mathbf{r}_2 = \sqrt{3}\cos\theta\,\mathbf{i}+\sin\theta\,\mathbf{j}$.

7 The motion of a particle at time t in seconds is given by

$$\mathbf{r} = 2\cos 4\pi t\,\mathbf{i}+\sin 4\pi t\,\mathbf{j}$$

Describe its motion.
A second particle follows this path in time t

$$\mathbf{r} = 2\cos 2\pi t\,\mathbf{j}-2\sin 2\pi t\,\mathbf{i}$$

If they both start when $t = 0$, at what times do they meet?

8 Find the distance between the intersections of

$$\mathbf{r} = 2\cos\phi\,\mathbf{i}+2\sin\phi\,\mathbf{j}+\cos\phi\,\mathbf{k}$$

and $\mathbf{r} = 2\cos\phi\,\mathbf{i}+2\sin\phi\,\mathbf{j}+(1-\cos\phi)\mathbf{k}$

9 Find the angle of inclination the major axis of the ellipse

$$\mathbf{r} = 4\cos\phi\,\mathbf{i}+3\sin\phi\,\mathbf{j}+3\cos\phi\,\mathbf{k}$$

makes with the plane $z = 0$. What is the eccentricity of the ellipse?

10 A beetle begins a walk on the locus

$$\mathbf{r} = 12\sin\frac{\pi}{4}t\,\mathbf{i}+5\cos\frac{\pi}{4}t\,\mathbf{j}+5\sin\frac{\pi}{4}t\,\mathbf{k}$$

when $t = 0$. At the highest point of its walk, on the first revolution, it slides down the major axis at $6\frac{1}{2}$ m s^{-1}. When it meets the curve again, it beetles back to its starting point on the same locus and at its former rate. What is its total time?

The Cycloid

A cycloid is the locus of a point on the circumference of a circle which rolls along a straight line.

Figure 3.14

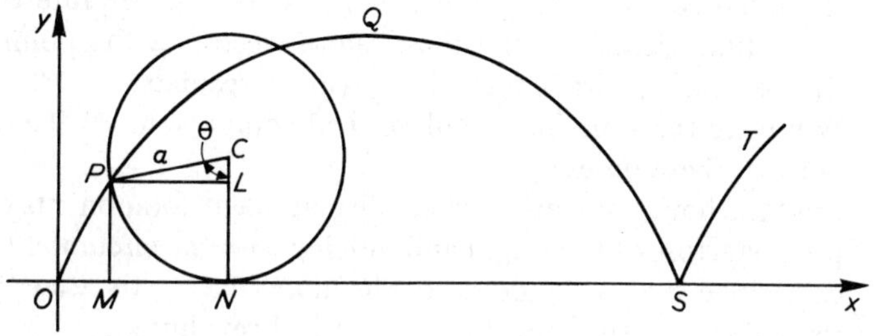

Figure 3.14 shows the cycloid $OPQST$. The point P is on the circumference of the circle, radius a, centre C. The parametric equations are formed by

$$\begin{aligned} x &= OM \\ &= ON - NM \\ &= \text{arc length } PN - LP \\ &= a\theta - a\sin\theta \end{aligned}$$

that is $\quad x = a(\theta - \sin\theta)$

Similarly $\quad y = MP$

$$= NC - CL = a - a\cos\theta = a(1-\cos\theta)$$

Therefore $\quad \mathbf{r} = x\mathbf{i} + y\mathbf{j}$

$$= a(\theta - \sin\theta)\mathbf{i} + a(1-\cos\theta)\mathbf{j}$$

For a circle of radius a, rolling along the x axis in the positive direction at an angular velocity ω, we have

$$\theta = \omega t \quad \text{and} \quad \mathbf{r} = a(\omega t - \sin\omega t)\mathbf{i} + a(1-\cos\omega t)\mathbf{j}$$

for any time $t \geqslant 0$.

Exercise 14

1 Plot the curves whose vector equations are
(i) $\mathbf{r} = 2(\theta - \sin\theta)\mathbf{i} + 2(1-\cos\theta)\mathbf{j}$
(ii) $\mathbf{r} = (p - \sin p)\mathbf{i} + (1-\cos p)\mathbf{j}$

2 A point on the circumference of a circle radius 3 units is at the origin when $t = 0$.
If the circle rolls along the positive x axis at the rate of $\frac{1}{2}$ revolution second^{-1}, find the position vector of the point at (i) 1 second (ii) $1\frac{1}{2}$ seconds (iii) $1\frac{3}{4}$ seconds.

3 What are the horizontal and vertical components of the distance at these times?

4 Find the horizontal and vertical displacement components of a point starting at the origin and moving with the circumference of a circle radius 1 unit as it rolls along the x axis after (i) $\frac{1}{4}$ revolution (ii) $\frac{1}{2}$ revolution (iii) 1 revolution.

5 For the cycloid $\mathbf{r} = (\theta - \sin\theta)\mathbf{i} + (1-\cos\theta)\mathbf{j}$, find the vector difference for the values (i) $\theta = \pi/6$ and (ii) $\theta = \pi/2$. What is the magnitude of this vector?

A non-planar curve: the helix

Lastly, we consider a curve in three dimensions

$$\mathbf{r} = x\mathbf{i} + y\mathbf{j} + z\mathbf{k}$$

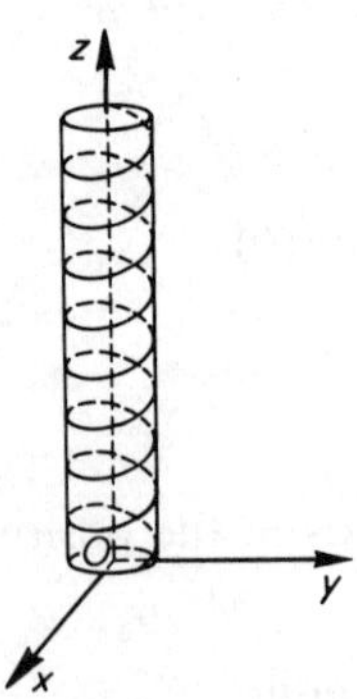

Figure 3.15

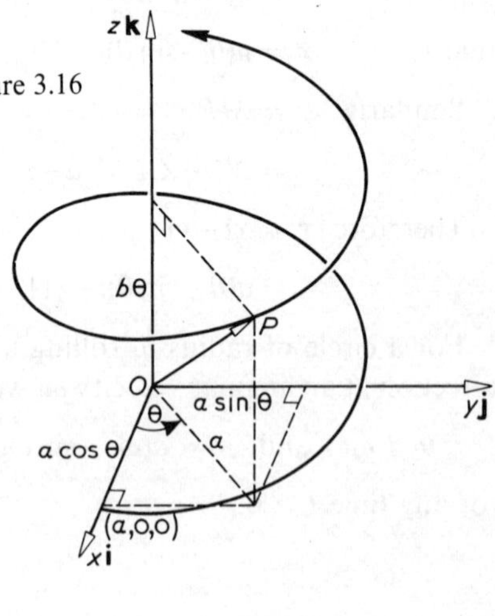

Figure 3.16

Figure 3.15 shows a circular helix winding its way around the z axis. It is the locus of a point drawn on the surface of a circular cylinder so that the curve cuts the generators of the cylinder at a constant angle. Many climbing plants follow the same route.

From Figure 3.16, we see that parametric equations for the curve are

$$x = a\cos\theta \qquad y = a\sin\theta \qquad z = b\theta$$

where a is the constant distance maintained by the point from the z axis, θ is the angle of revolution and b is the constant required to determine the position of P relative to the z axis for a given angle of revolution.

Hence $\mathbf{r} = a\cos\theta\,\mathbf{i} + a\sin\theta\,\mathbf{j} + b\theta\,\mathbf{k}$

Again, in terms of t measured in time, we can have

$$\mathbf{r} = a\cos\omega t\,\mathbf{i} + a\sin\omega t\,\mathbf{j} + b\omega t\,\mathbf{k}$$

where ω is the angular velocity of the point around the z axis.

When the distance described by the point from the z axis is not constant, the parametric equations are

$$x = a\cos\theta \qquad y = b\sin\theta \qquad z = c\theta$$

The curve is now no longer a circular helix but an elliptic helix.

Example

A point on a circular helix starts at (0, 3, 0) when $t = 0$. When $t = 2$ seconds the point is at $(0, 3, 6\pi)$. If the angular velocity of the point is π radians second^{-1}, find the vector position of the point for any time t.

$\theta = \pi t$, and when $t = 2$, we have

$$b\theta = 2\pi b = 6\pi$$

that is $\qquad b = 3$

Hence $\mathbf{r} = 3\cos\pi t\,\mathbf{i} + 3\sin\pi t\,\mathbf{j} + 3\pi t\,\mathbf{k}$

Summary

Equations of a plane

$$(\mathbf{q}-\mathbf{p}).\mathbf{n} = 0$$

$$\mathbf{q}.\mathbf{n} = c$$

or $\quad q_x x + q_y y + q_z z = c$

where q_x, q_y and q_z are the components of the vector perpendicular to the plane and c is a number determined by the dot product of the vector of a point in the plane and $\mathbf{n}$, $\mathbf{p}.\mathbf{n}$.

Perpendicular distance from a point to a plane
(i) When the point is at the origin

$$d = |\mathbf{p}.\hat{\mathbf{n}}|$$

where $\hat{\mathbf{n}}$ is the unit normal to the plane and $\mathbf{p}$ is the vector of a point in the plane.
(ii) When the point is not at the origin

$$d = |\mathbf{p}.\hat{\mathbf{n}} - \mathbf{s}.\hat{\mathbf{n}}| = |(\mathbf{p}-\mathbf{s}).\hat{\mathbf{n}}|$$

where $\mathbf{s}$ is the vector of the point on the plane parallel to the given plane.

Angle between two planes

$$\cos\theta = \frac{\mathbf{n}_1.\mathbf{n}_2}{n_1 n_2} = \hat{\mathbf{n}}_1.\hat{\mathbf{n}}_2$$

Equations of a straight line
The equation of a line through the point P whose vector is $\mathbf{p}$ and parallel to a given vector $\mathbf{q}$ is

$$\mathbf{r} = \mathbf{p} + t\mathbf{q}$$

where t is a real number to define a particular vector, $\mathbf{r}$.
For parameter t, we have

$$\frac{x-p_x}{q_x} = \frac{y-p_y}{q_y} = \frac{z-p_z}{q_z} = t$$

For a line through the points P and Q, vectors $\mathbf{p}$ and $\mathbf{q}$, we have

$$\mathbf{r} = \mathbf{p} + t(\mathbf{q}-\mathbf{p})$$

For two non-parallel planes with orthogonal vectors **q** and **s**, the line of intersection is

$$\mathbf{r} = \mathbf{p}+t(\mathbf{q} \times \mathbf{s})$$

where **p** is the vector of a point P on the line.

Equations of a circle

$$\mathbf{r} = a\cos\theta\,\mathbf{i}+a\sin\theta\,\mathbf{j}$$

where a is the radius of the circle.

$$\mathbf{r} = a\cos\left(\frac{s}{a}\right)\mathbf{i}+a\sin\left(\frac{s}{a}\right)\mathbf{j}$$

where s is the arc length in the range $0 \leqslant s \leqslant 2\pi a$.

$$\mathbf{r} = a\cos\omega t\,\mathbf{i}+a\sin\omega t\,\mathbf{j}$$

where ω is the angular velocity of the radius vector and t is in units of time.

Equation for a parabola

$$\mathbf{r} = ap^2\mathbf{i}+2ap\mathbf{j}$$

Equation for an ellipse

$$\mathbf{r} = a\cos\phi\,\mathbf{i}+b\sin\phi\,\mathbf{j}$$

Equation for a hyperbola

$$\mathbf{r} = a\sec\phi\,\mathbf{i}+b\tan\phi\,\mathbf{j}$$

Equation for a cycloid

$$\mathbf{r} = a(\theta-\sin\theta)\mathbf{i}+a(1-\cos\theta)\mathbf{j}$$

Equation for a circular helix

$$\mathbf{r} = a\cos\theta\,\mathbf{i}+a\sin\theta\,\mathbf{j}+b\theta\,\mathbf{k}$$

where a is the constant distance from the z axis.

Equation for an elliptic helix

$$\mathbf{r} = a\cos\theta\,\mathbf{i}+b\sin\theta\,\mathbf{j}+c\theta\,\mathbf{k}$$

Exercise 15

1 Sketch the curves whose vector equations are
(i) $\mathbf{r} = 3\cos\phi\,\mathbf{i}+3\sin\phi\,\mathbf{j}+2\phi\,\mathbf{k}$
(ii) $\mathbf{r} = 3\cos\phi\,\mathbf{i}+5\sin\phi\,\mathbf{j}+2\phi\,\mathbf{k}$
(iii) $\mathbf{r} = a\cos p\,\mathbf{i}+a\sin p\,\mathbf{j}+bp\,\mathbf{k}$

2 For the circular helix $\mathbf{r} = a\cos\theta\,\mathbf{i}+a\sin\theta\,\mathbf{j}+b\theta\,\mathbf{k}$ a point starts from (4, 0, 0) when $t = 0$ and is at (4, 0, π) when $t = 1$. The point rotates about the z axis at 5 revolutions second^{-1}. Find the position vector equation of motion with respect to time.

3 Find the condition that the line $\mathbf{r}_1 = t(4\mathbf{i}+3\mathbf{j}+\mathbf{k})$ intersects the helix $\mathbf{r}_2 = a\cos p\,\mathbf{i}+a\sin p\,\mathbf{j}+bp\,\mathbf{k}$

4 Find values of ϕ for which the helices
$\mathbf{r}_1 = a\cos\phi\,\mathbf{i}+a\sin\phi\,\mathbf{j}+a\phi\mathbf{k}$
$\mathbf{r}_2 = a\cos 2\phi\,\mathbf{i}+a\sin 2\phi\,\mathbf{j}+a\delta\mathbf{k}$
coincide.

5 What is the length of the segment between the first two intersections of the helices of question 4?

6 Find the intersections of the circles
$\mathbf{r}_1 = \cos p\,\mathbf{i}+\sin p\,\mathbf{j}$
$\mathbf{r}_2 = \cos p\,\mathbf{i}+\sin p\,\mathbf{j}+\mathbf{k}$
with the helix $\mathbf{r}_3 = a\cos p\,\mathbf{i}+a\sin p\,\mathbf{j}+bp\,\mathbf{k}$.

7 What is the distance of the point P from the origin where P is a point on the helix $\mathbf{r} = \cos p\,\mathbf{i}-4\sin p\,\mathbf{j}+5p\,\mathbf{k}$ when $p = 3\pi/2$?

8 Plot the curve whose vector equation is $\mathbf{r} = 2ap^2\mathbf{i}+ap^3\mathbf{j}$.

9 Plot the curve whose vector equation is $\mathbf{r} = 3\cos p\,\mathbf{i}+2\sin p\,\mathbf{j}$.

10 On the surface of the cylinder determined by the helix described in question 2, other points generate circles parallel to the x, y plane at times $t = 0, 1, 2, 3, \ldots$ seconds. Write down the equation of the circle when $t = 1$ second, find the distance between two successive circles and form the equation of the straight line joining the initial point of the helix to the point determined by the position vector of the helix when $t = 2\frac{1}{4}$ seconds.

4 · *Differentiation and integration*

Differentiation

We next consider the behaviour of vectors as they change with respect to time.

Suppose that $\mathbf{r}$ is a vector which varies continuously in magnitude and direction. If $\mathbf{r}$ is a continuous and single-valued function of a scalar variable t, then when t increases by a small quantity δt, $\mathbf{r}$ changes correspondingly by a small vector quantity $\delta\mathbf{r}$.

Figure 4.1

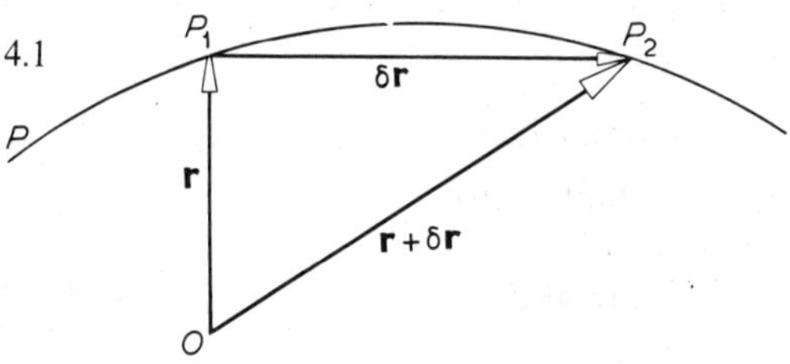

Hence, if P is a point moving along a curve (Figure 4.1), and t changes to $t+\delta t$, then we suppose that $\mathbf{r}$ changes to $\mathbf{r}+\delta\mathbf{r}$ and the point changes from its position P_1 to its new position P_2.

The ratio $\delta\mathbf{r}/\delta t$ is the average rate of change of $\mathbf{r}$ with respect to t and as δt tends to zero, we obtain the limiting value $d\mathbf{r}/dt$. This is the rate of increase of $\mathbf{r}$ with respect to t

$$\frac{d\mathbf{r}}{dt} = \lim_{\delta t \to 0} \frac{\delta\mathbf{r}}{\delta t}$$

$d\mathbf{r}/dt$ is called the derivative of vector $\mathbf{r}$ with respect to scalar t. As δt tends to zero, the direction of P_1P_2 tends to the direction of the tangent at P_1. Therefore the direction of $d\mathbf{r}/dt$ is along the tangent at P_1 (Figure 4.2).

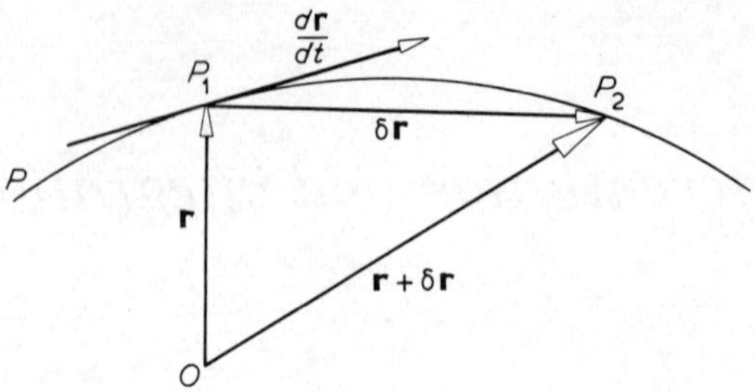

Figure 4.2

When $d\mathbf{r}/dt$ is itself a continuous and single valued function of t then we can similarly obtain the second derivative $d^2\mathbf{r}/dt^2$. Further, we may win higher order derivatives where they exist.

If, now, $\mathbf{r}$ represents the displacement of a particle and t is the scalar representing time, then $\delta\mathbf{r}/\delta t$ represents the average velocity of the particle during the interval δt and $d\mathbf{r}/dt$, or $\dot{\mathbf{r}}$ represents the velocity of the particle along the path at any instant.

Similarly, $\delta^2\mathbf{r}/\delta t^2$ represents the average acceleration during the interval δt and $d^2\mathbf{r}/dt^2$ of $\ddot{\mathbf{r}}$ represents the corresponding instantaneous acceleration.

Derivative of a constant vector

Let $\mathbf{r} = \mathbf{i}$, a constant vector.

Then $\mathbf{r}+\delta\mathbf{r} = \mathbf{i}+\delta\mathbf{i}$

but $\delta\mathbf{i} = \mathbf{0}$. Therefore $\delta\mathbf{i}/\delta t = \mathbf{0}$. Hence, the derivative of a constant vector is the zero vector.

Derivative of a sum

Let $\mathbf{r} = \mathbf{p}+\mathbf{q}$, then

$$\mathbf{r}+\delta\mathbf{r} = \mathbf{p}+\delta\mathbf{p}+\mathbf{q}+\delta\mathbf{q}$$

that is $$\delta\mathbf{r} = \delta\mathbf{p}+\delta\mathbf{q}$$

or $$\frac{\delta\mathbf{r}}{\delta t} = \frac{\delta\mathbf{p}}{\delta t}+\frac{\delta\mathbf{q}}{\delta t}$$

As δt tends to zero, we have

$$\lim_{\delta t\to 0}\frac{\delta\mathbf{r}}{\delta t} = \frac{d\mathbf{r}}{dt} = \frac{d\mathbf{p}}{dt}+\frac{d\mathbf{q}}{dt}$$

This process may be extended to include any number of vectors.

Derivative of a product

Case 1. *Product of a constant coefficient and vector*
Let $\mathbf{r} = a\mathbf{p}$, where a is a constant coefficient, then

$$\mathbf{r}+\delta\mathbf{r} = a(\mathbf{p}+\delta\mathbf{p})$$
$$\delta\mathbf{r} = a\,\delta\mathbf{p}$$
$$\frac{\delta\mathbf{r}}{\delta t} = a\frac{\delta\mathbf{p}}{\delta t}$$

Therefore

$$\lim_{\delta t\to 0}\frac{\delta\mathbf{r}}{\delta t} = \frac{d\mathbf{r}}{dt} = a\,\frac{d\mathbf{p}}{dt}$$

Case 2. *Product of a parameter and vector.*
Let $\mathbf{r} = u\mathbf{p}$, where u is a parameter.

$$\mathbf{r}+\delta\mathbf{r} = (u+\delta u)(\mathbf{p}+\delta\mathbf{p})$$
$$= u\mathbf{p}+u\,\delta\mathbf{p}+\delta u\,\mathbf{p}+\delta u\,.\,\delta\mathbf{p}$$
$$\frac{\delta\mathbf{r}}{\delta t} = u\frac{\delta\mathbf{r}}{\delta t}+\frac{\delta u}{\delta t}\mathbf{p}+\frac{\delta u\,.\,\delta\mathbf{p}}{\delta t}$$

Therefore

$$\frac{d\mathbf{r}}{dt} = u\frac{d\mathbf{p}}{dt}+\frac{du}{dt}\mathbf{p}$$

when we neglect the second order term.

Case 3. *Dot product of two vectors.*
Let $\mathbf{r} = \mathbf{p}\,.\,\mathbf{q}$, then

$$\mathbf{r}+\delta\mathbf{r} = (\mathbf{p}+\delta\mathbf{p})\,.\,(\mathbf{q}+\delta\mathbf{q})$$
$$= \mathbf{p}\,.\,\mathbf{q}+\mathbf{p}\,.\,\delta\mathbf{q}+\delta\mathbf{p}\,.\,\mathbf{q}+\delta\mathbf{p}\,.\,\delta\mathbf{q}$$
$$\frac{\delta\mathbf{r}}{\delta t} = \mathbf{p}\frac{\delta\mathbf{q}}{\delta t}+\frac{\delta\mathbf{p}}{\delta t}\mathbf{q}+\frac{\delta\mathbf{p}\,.\,\delta\mathbf{q}}{\delta t}$$

Therefore

$$\frac{d\mathbf{r}}{dt} = \mathbf{p}\cdot\frac{d\mathbf{q}}{dt}+\frac{d\mathbf{p}}{dt}\cdot\mathbf{q}$$

when we neglect the second order term.

Case 4. *Cross product of two vectors.*

$$\mathbf{r} = \mathbf{p}\times\mathbf{q}$$

$$\mathbf{r}+\delta\mathbf{r} = (\mathbf{p}+\delta\mathbf{p})\times(\mathbf{q}+\delta\mathbf{q})$$

$$= \mathbf{p}\times\mathbf{q}+\mathbf{p}\times\delta\mathbf{q}+\delta\mathbf{p}\times\mathbf{q}+\delta\mathbf{p}\times\delta\mathbf{q}$$

$$\delta\mathbf{r} = \mathbf{p}\times\delta\mathbf{q}+\delta\mathbf{p}\times\mathbf{q}+\delta\mathbf{p}\times\delta\mathbf{q}$$

Therefore

$$\frac{d\mathbf{r}}{dt} = \mathbf{p}\times\frac{d\mathbf{q}}{dt}+\frac{d\mathbf{p}}{dt}\times\mathbf{q}$$

when we neglect the second order term.

These differentiation rules are the same as those in the ordinary differential calculus with the notable exception of the cross product. The order of the differentiation here must be strictly kept because of its non-commutative property.

When x, y and z are the components of $\mathbf{r}$ in the direction of the coordinate axes, we have

$$\mathbf{r} = x\mathbf{i}+y\mathbf{j}+z\mathbf{k}$$

and applying these rules, we have

$$\frac{d\mathbf{r}}{dt} = \frac{dx}{dt}\mathbf{i}+\frac{dy}{dt}\mathbf{j}+\frac{dz}{dt}\mathbf{k}$$

and $$\frac{d^2\mathbf{r}}{dt^2} = \frac{d^2x}{dt^2}\mathbf{i}+\frac{d^2y}{dt^2}\mathbf{j}+\frac{d^2z}{dt^2}\mathbf{k}$$

Example 1

Suppose in Case 3, we have $\mathbf{q} = \mathbf{p}$, and we find the derivative of $\mathbf{p}.\mathbf{p}$, that is $\mathbf{r} = \mathbf{p}^2$.

We know

$$\frac{d\mathbf{r}}{dt} = \left(\frac{d}{dt}\right)\mathbf{p}^2$$

$$= \left(\frac{d}{dt}\right)p^2$$

That is, $2p\dfrac{dp}{dt} = 2\mathbf{p}.\dfrac{d\mathbf{p}}{dt}$

Hence $\mathbf{p}.\frac{d\mathbf{p}}{dt} = p\frac{dp}{dt}$

When $\mathbf{p}$ is of constant length, then $\mathbf{p}^2$ is constant and

$$\mathbf{p}.\frac{d\mathbf{p}}{dt} = 0$$

Therefore $d\mathbf{p}/dt$ is perpendicular to $\mathbf{p}$, that is, the derivative of a constant length vector is perpendicular to the vector. We use this property to find the normal to a curve on page 101.

Example 2

A point moves along a curve such that after time t

$$\mathbf{r} = (2\sin t - \cos t)\mathbf{i} + (3\sin t + t^2)\mathbf{j} + e^{4t}\mathbf{k}$$

Find the velocity and acceleration of the particle after $\pi/2$ seconds.

$$\dot{\mathbf{r}} = (2\cos t + \sin t)\mathbf{i} + (3\cos t + 2t)\mathbf{j} + 4e^{4t}\mathbf{k}$$

$$\dot{\mathbf{r}}_{t=\pi/2} = \mathbf{i} + \pi\mathbf{j} + 4e^{2\pi}\mathbf{k}$$

$$\ddot{\mathbf{r}} = (-2\sin t + \cos t)\mathbf{i} + (-3\sin t + 2)\mathbf{j} + 16e^{4t}\mathbf{k}$$

$$\ddot{\mathbf{r}}_{t=\pi/2} = -2\mathbf{i} + (-1)\mathbf{j} + 16e^{2\pi}\mathbf{k}$$

$$= -2\mathbf{i} - \mathbf{j} + 16e^{2\pi}\mathbf{k}$$

Tangents

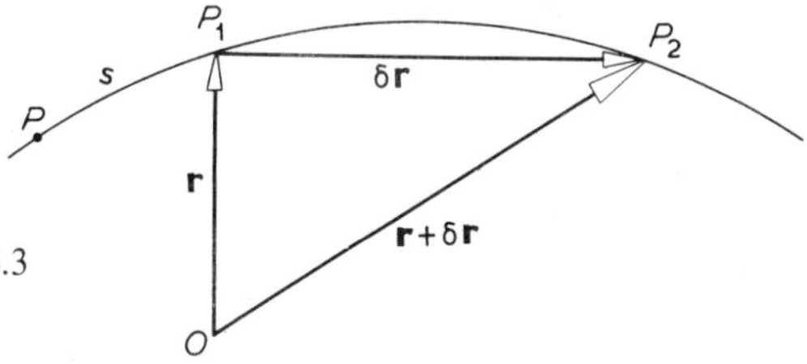

Figure 4.3

Suppose s is the length PP_1 of the arc traced out by a moving point from a fixed point P on a curve (Figure 4.3).

Let $\mathbf{r} = \mathbf{OP}_1$, $\mathbf{r} + \delta\mathbf{r} = \mathbf{OP}_2$, $\delta\mathbf{r} = \mathbf{P}_1\mathbf{P}_2$ and the arc length $P_1P_2 = \delta s$.

Now $\delta\mathbf{r}/\delta s$ is the ratio of the chord $\mathbf{P_1P_2}$ to the length of the arc P_1P_2 in the direction of the vector $\delta\mathbf{r}$.

Further the magnitude of the quotient, that is $|\delta\mathbf{r}/\delta s|$, tends to unity as δs tends to zero, since the ratio of the lengths, chord P_1P_2 to arc P_1P_2, tends to unity as P_2 tends to P_1.

Therefore if $\hat{\mathbf{t}}$, say, is the limit of $\delta\mathbf{r}/\delta s$ as δs tends to zero, we have

$$\lim_{\delta s \to 0} \frac{\delta\mathbf{r}}{\delta s} = \frac{d\mathbf{r}}{ds} = \hat{\mathbf{t}}$$

Now $\hat{\mathbf{t}}$ is a unit vector in the direction of the tangent to the curve at P_1: we refer to $\hat{\mathbf{t}}$ as the unit tangent.

When $\quad \mathbf{r} = x\mathbf{i}+y\mathbf{j}+z\mathbf{k}$

then $\quad \dfrac{d\mathbf{r}}{ds} = \dfrac{dx}{ds}\mathbf{i}+\dfrac{dy}{ds}\mathbf{j}+\dfrac{dz}{ds}\mathbf{k}$

The values $\dfrac{dx}{ds}, \dfrac{dy}{ds}$ and $\dfrac{dz}{ds}$ are the direction cosines of $\hat{\mathbf{t}}$.

So what is the equation of the tangent to the curve at P_1?

Figure 4.4

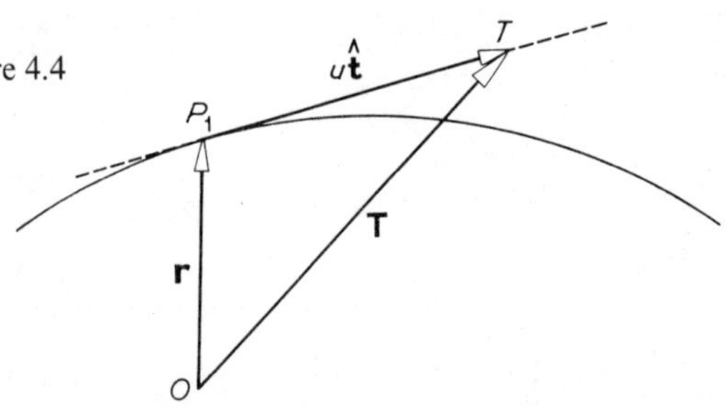

If T is any point on the tangent (Figure 4.4), then the position vector of T is

$$\mathbf{T} = \mathbf{r}+u\hat{\mathbf{t}}$$

for parameter u.

Normals

Using the result of the first example on page 98,

since $\quad \hat{\mathbf{t}}.\hat{\mathbf{t}} = 1 \quad \text{and} \quad \hat{\mathbf{t}}.\dfrac{d\hat{\mathbf{t}}}{ds} = 0$

we have $\quad \dfrac{d\hat{\mathbf{t}}}{ds} = \dfrac{d^2\mathbf{r}}{ds^2} = \dfrac{\hat{\mathbf{n}}}{\rho}$

where $\hat{\mathbf{n}}$ is a unit vector parallel to $d\hat{\mathbf{t}}/ds$ and called the principal normal and ρ is a scalar quantity called the radius of curvature, which we discuss before the next example.

For a positive vector **N**, therefore, on the principal normal, the equation is

$$\mathbf{N} = \mathbf{r} + u\hat{\mathbf{n}}$$

for parameter u.

Radius of curvature

Figure 4.5

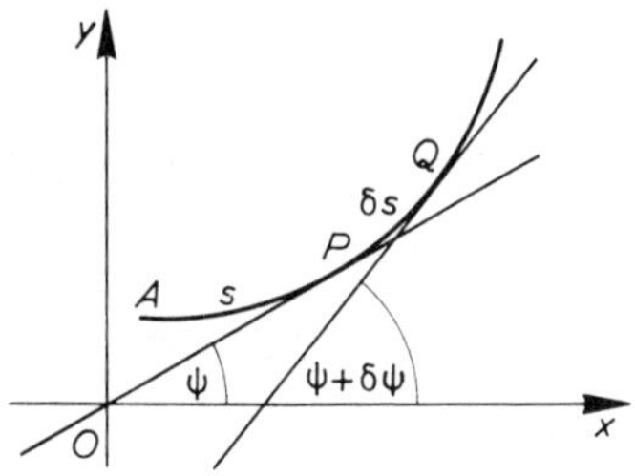

Let two points on a curve be P and Q so that $AP = s$ and $AQ = s + \delta s$ (Figure 4.5). Let the tangents at P and Q make angles of ψ and $\psi + \delta\psi$ respectively with the positive direction of the x axis.

The ratio $\delta\psi/\delta s$ is the average rate of change of ψ with respect to s between the fixed points.

As Q gets closer to P, we approach the limit of this quotient until

$$\lim_{\delta s \to 0} \frac{\delta\psi}{\delta s} = \frac{d\psi}{ds}$$

This quantity is called the curvature of the curve at P.

In particular, when the curve is a circle, we have

$\dfrac{d\psi}{ds} = \dfrac{1}{r}$ where r is the radius.

Figure 4.6

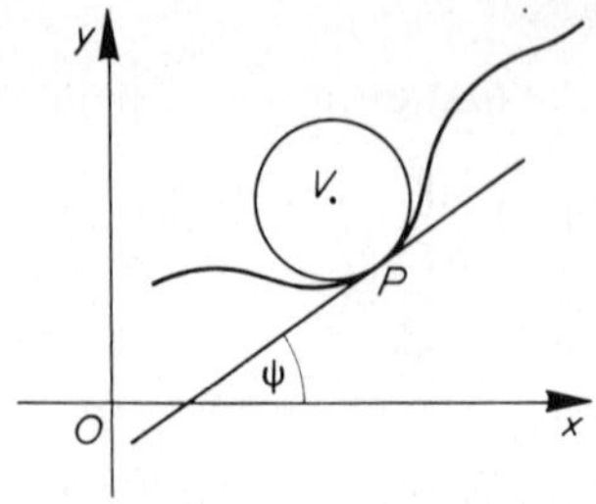

Any point on a curve which is not a circle can have a circle touching the curve at the point P (Figure 4.6).

This circle has the same curvature as the curve at P. Its radius is the radius of curvature of the curve at P and its centre V is the centre of curvature of the curve at P.

If ρ is the radius of curvature at P, then

$$\frac{1}{\rho} = \frac{d\psi}{ds} \quad \text{or} \quad \rho = \frac{ds}{d\psi}.$$

Example

Find the unit tangents and normals to the circular helix

$$\mathbf{r} = a\cos\theta\,\mathbf{i} + a\sin\theta\,\mathbf{j} + b\theta\,\mathbf{k}$$

$$\hat{\mathbf{t}} = \frac{d\mathbf{r}}{ds} = -a\sin\theta\frac{d\theta}{ds}\mathbf{i} + a\cos\theta\frac{d\theta}{ds}\mathbf{j} + b\frac{d\theta}{ds}\mathbf{k}$$

$$= a\frac{d\theta}{ds}\left(-\sin\theta\,\mathbf{i} + \cos\theta\,\mathbf{j} + \frac{b}{a}\mathbf{k}\right)$$

Since $\hat{\mathbf{t}}$ is a unit vector, we have

$$a^2\left(\frac{d\theta}{ds}\right)^2\left(\sin^2\theta + \cos^2\theta + \frac{b^2}{a^2}\right) = 1$$

that is $a^2\left(\frac{d\theta}{ds}\right)^2\left(1 + \frac{b^2}{a^2}\right) = 1$

whence $\dfrac{d\theta}{ds} = \dfrac{1}{\sqrt{a^2+b^2}}$

Therefore $\hat{\mathbf{t}} = \dfrac{a}{\sqrt{a^2+b^2}}\left(-\sin\theta\,\mathbf{i} + \cos\theta\,\mathbf{j} + \dfrac{b}{a}\mathbf{k}\right)$

Again $$\frac{\hat{\mathbf{n}}}{\rho} = \frac{d\hat{\mathbf{t}}}{ds} = \frac{d^2\mathbf{r}}{ds^2} = \frac{a}{\sqrt{a^2+b^2}}\left(-\cos\theta\,\mathbf{i}-\sin\theta\,\mathbf{j}\right)\frac{d\theta}{ds}$$

or $$= \frac{a}{a^2+b^2}\left(-\cos\theta\,\mathbf{i}-\sin\theta\,\mathbf{j}\right)$$

$$\hat{\mathbf{n}} = -\frac{\rho a}{a^2+b^2}\left(\cos\theta\mathbf{i}+\sin\theta\mathbf{j}\right)$$

For this unit vector, therefore, we have

$$|\hat{\mathbf{n}}|^2 = \frac{\rho^2 a^2}{(a^2+b^2)^2}\left(\cos^2\theta+\sin^2\theta\right) = 1$$

whence $$\rho = \frac{a^2+b^2}{a}$$

Radial and transverse components of velocity and acceleration

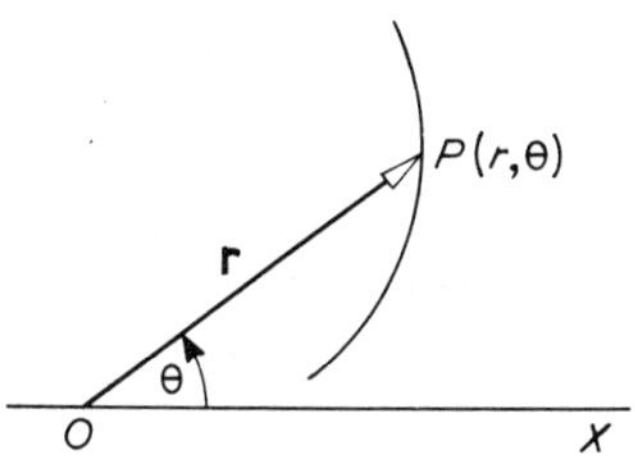

Figure 4.7

In Figure 4.7 the point P moves along the path of a curve in the plane of the paper. The origin is O and θ is the inclination of **OP**, vector **r**, to a fixed line of reference OX. The polar coordinates of P are then (r, θ).

We now obtain the radial velocity, that is the velocity parallel to **r**, and the transverse velocity perpendicular to **r**.
Since **r** is only in two dimensions, we have

$$\mathbf{r} = x\mathbf{i}+y\mathbf{j} = r\cos\theta\,\mathbf{i}+r\sin\theta\,\mathbf{j}$$

Therefore

$$\dot{\mathbf{r}} = (\dot{r}\cos\theta-r\sin\theta\,\dot{\theta})\,\mathbf{i}+\ (\dot{r}\sin\theta+r\cos\theta\,\dot{\theta})\,\mathbf{j}$$

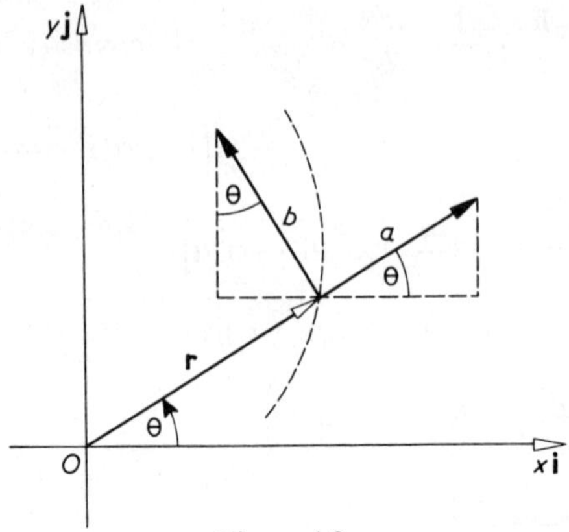

Figure 4.8

In Figure 4.8, a and b are the radial and transverse components respectively of the velocity of P. From the figure, we see that

$$\dot{\mathbf{r}} = (a\cos\theta - b\sin\theta)\mathbf{i} + (a\sin\theta + b\cos\theta)\mathbf{j}$$

Comparing the respective components of **i** and **j**, we have

$$a = \dot{r} \quad \text{and} \quad b = r\dot{\theta}$$

that is the radial velocity is $\dot{\mathbf{r}}$ and the transverse velocity is $\mathbf{r}\dot{\theta}$.

We can write the velocity vector, therefore, as

$$\dot{\mathbf{r}} = \dot{r}\hat{\mathbf{r}} + r\dot{\theta}\hat{\mathbf{q}}$$

where $\hat{\mathbf{r}}$ is a unit vector in the direction of the radius and $\hat{\mathbf{q}}$ is a unit vector perpendicular to the radius, in the transverse direction. As we can see from the Figures 4.7 and 4.8 $\hat{\mathbf{q}}$ is not in general in the direction of the tangent at P.

Again

$$\begin{aligned}\ddot{\mathbf{r}} &= (\ddot{r}\cos\theta - \dot{r}\sin\theta\,\dot{\theta} - \dot{r}\sin\theta\,\dot{\theta} - r\cos\theta\,\dot{\theta}^2 - r\sin\theta\,\ddot{\theta})\mathbf{i}\\ &\quad + (\ddot{r}\sin\theta + \dot{r}\cos\theta\,\dot{\theta} + \dot{r}\cos\theta\,\dot{\theta} - r\sin\theta\,\dot{\theta}^2 + r\cos\theta\,\ddot{\theta})\mathbf{j}\\ &= \{(\ddot{r} - r\dot{\theta}^2)\cos\theta - (2\dot{r}\dot{\theta} + r\ddot{\theta})\sin\theta\}\mathbf{i}\\ &\quad + \{(\dot{r} - r\dot{\theta}^2)\sin\theta + (2\dot{r}\dot{\theta} + r\ddot{\theta})\cos\theta\}\mathbf{j}\end{aligned}$$

If we now let c and d be the radial and transverse components respectively of the acceleration of P, we have

$$c = \ddot{r} - r\dot{\theta}^2 \quad \text{and} \quad d = 2\dot{r}\dot{\theta} + r\ddot{\theta}$$

$$= \frac{1}{r}(2r\dot{r}\dot{\theta}+r^2\ddot{\theta})$$

$$= \frac{1}{r}\frac{d}{dt}\left(r^2\dot{\theta}\right)$$

Again we can write the acceleration vector in a similar form to the velocity vector

$$\ddot{\mathbf{r}} = (\ddot{r}-r\dot{\theta}^2)\hat{\mathbf{r}}+(2\dot{r}\dot{\theta}+r\ddot{\theta})\hat{\mathbf{q}}$$

Example

The velocities of a particle along and perpendicular to a radius vector from a fixed origin are λr^2 and $\mu\theta^2$; find the component accelerations in terms of r and θ. (*Part of CS question*)

Since $\dot{r} = \lambda r^2$
we have $\ddot{r} = 2\lambda r\dot{r}$
$= 2\lambda^2 r^3$
Again $r\dot{\theta} = \mu\theta^2$
hence $\dot{\theta} = \mu\theta^2/r$
and $\dot{\theta}^2 = \mu^2\theta^4/r^2$
Therefore $r\dot{\theta}^2 = \mu^2\theta^4/r$.

We have, therefore, the radial acceleration,

$$\ddot{r}-r\,\dot{\theta}^2 = 2\lambda^2 r^3-\mu^2\theta^4/r$$

Transverse acceleration $= 2\dot{r}\dot{\theta}+r\ddot{\theta}$

$$= \frac{1}{r}\frac{d}{dt}\left(r^2\dot{\theta}\right)$$

$$= \frac{1}{r}\frac{d}{dt}\left(r^2\frac{\mu\theta^2}{r}\right)$$

$$= \frac{1}{r}\frac{d}{dt}\left(\mu r\theta^2\right)$$

$$= \frac{1}{r}\left(\mu\theta^2\dot{r}+2\mu r\theta\dot{\theta}\right)$$

$$= \frac{1}{r}\left(\mu\theta^2\lambda r^2+\frac{2\mu r\theta\,\mu\theta^2}{r}\right)$$

Therefore the transverse acceleration is $\mu\theta^2\left(\lambda r+\frac{2\mu\theta}{r}\right)$

Normal and tangential components of acceleration

We know that the velocity of P at any instant is given by $\dot{\mathbf{r}} = \mathbf{v}$, say. This may also be written

$$\dot{\mathbf{r}} = \mathbf{v} = \frac{d\mathbf{r}}{ds}\frac{ds}{dt} = v\hat{\mathbf{t}}$$

where $v = ds/dt$ is the speed of the particle along the path and $\hat{\mathbf{t}}$ is the unit tangent to the curve at the point P. (Guard against confusion here between $\hat{\mathbf{t}}$, the unit tangent, and t, the scalar time.)

To find the acceleration, we differentiate $v\hat{\mathbf{t}}$ with respect to time, t, that is

$$\begin{aligned}\ddot{\mathbf{r}} = \mathbf{a} &= \frac{dv}{dt}\hat{\mathbf{t}} + v\frac{d\hat{\mathbf{t}}}{dt}\\ &= \frac{dv}{dt}\hat{\mathbf{t}} + v\frac{d\hat{\mathbf{t}}}{ds}\frac{ds}{dt}\\ &= \frac{dv}{dt}\hat{\mathbf{t}} + v\frac{\hat{\mathbf{n}}}{\rho}v\\ &= \frac{dv}{dt}\hat{\mathbf{t}} + \frac{v^2}{\rho}\hat{\mathbf{n}}\end{aligned}$$

Therefore the acceleration of a point moving along a curve may be resolved into a component dv/dt along the tangent and a component v^2/ρ along the principal normal (Figure 4.9).

Figure 4.9

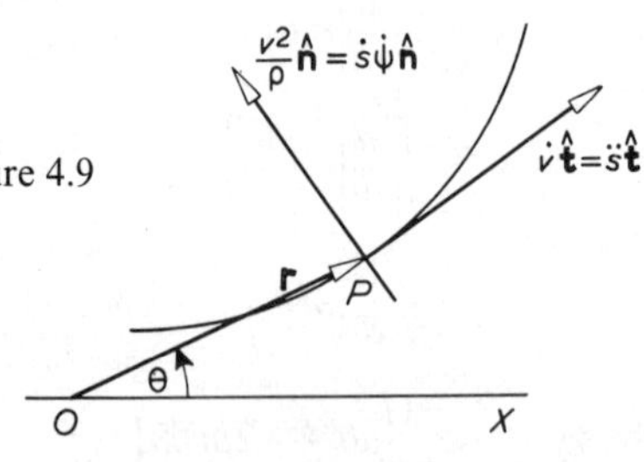

Since $\quad \dfrac{dv}{dt} = \ddot{s} \quad$ and $\quad \dfrac{v^2}{\rho} = \dfrac{\dot{s}^2}{\rho} = \dot{s}^2\dfrac{d\psi}{ds} = \dot{s}^2\dfrac{d\psi}{dt}\dfrac{dt}{ds} = \dot{s}\dot{\psi},$

we have another way to express this acceleration

$$\ddot{\mathbf{r}} = \ddot{s}\hat{\mathbf{t}} + \dot{s}\dot{\psi}\hat{\mathbf{n}}$$

Now $\dot{\psi}$ is the rate at which the unit tangent to the curve rotates.

Example

A particle moves on the curve $s = c \tan \psi$ in such a way that the tangent to the curve rotates uniformly. Show that the tangential acceleration varies jointly as the radius of curvature and the path, and that the normal acceleration is proportional to the radius of curvature.

$$s = c \tan \psi \qquad \ldots \qquad \ldots \text{I}$$

$$\dot{s} = c \sec^2\psi \, \dot{\psi}$$

but $\dot{\psi}$ is constant, therefore

$$\dot{s} = a \sec^2 \psi \qquad \ldots \qquad \ldots \text{II}$$

for some constant a.

Again

$$\ddot{s} = 2a \sec^2\psi \tan \psi \, \dot{\psi}$$

$$= bs \sec^2\psi \quad \text{(substituting from I)}$$

where b is a constant.

For the radius of curvature, we have

$$\rho = \frac{ds}{d\psi} = c \sec^2\psi \qquad \ldots \qquad \ldots \text{III}$$

Therefore we have the tangential acceleration

$$\ddot{s} = A\rho s$$

which varies as the radius of curvature ρ and the path s.

The normal acceleration is given by

$$\dot{s}\dot{\psi} = B\dot{s}$$

because the tangent rotates uniformly. B is a constant.

Hence $\dot{s}\dot{\psi} = Ba \sec^2\psi$ (from II)

$= D\rho$ (from III)

for a constant D, and so the normal acceleration is proportional to ρ.

Integration

If a certain vector function $\mathbf{r}$ of a scalar variable t is given, then the process of finding another function $\mathbf{F}$ of t such that

$$\frac{d\mathbf{F}}{dt} = \mathbf{r}$$

is called integration.

Integration is, therefore, the inverse of differentiation. When $\mathbf{F}$ exists, we write the integral of $\mathbf{r}$ with respect to t, as

$$\mathbf{F} = \int \mathbf{r}\, dt + \mathbf{c}$$

This is equivalent to

$$\frac{d\mathbf{F}}{dt} = \mathbf{r}$$

As in the scalar calculus, $\mathbf{F}$ is known as the *indefinite integral* since the derivative of $\mathbf{F}+\mathbf{c}$, where $\mathbf{c}$ is a constant vector, is also $\mathbf{F}$.

The constant of integration $\mathbf{c}$ is usually found either from initial conditions set by the problem or from defined geometrical properties of the curve.

The *definite integral* is

$$\int_{t_1}^{t_2} \mathbf{r}\, dt = \mathbf{F}(t_2) - \mathbf{F}(t_1)$$

Example

A particle is projected from ground level with a velocity of 19·62 m s^{-1} at an elevation of 30°. Find (i) the time taken for the particle to reach the horizontal plane through the point of projection, (ii) the horizontal range and (iii) the greatest height of the particle. We assume that the particle moves freely under the action of gravity which is approximately 9·81 m s^{-2}. Let $\mathbf{i}$ and $\mathbf{j}$ be unit vectors horizontally and vertically respectively.

Since there is no horizontal acceleration, we have

$$\ddot{\mathbf{r}} = -9{\cdot}81\mathbf{j}$$

Therefore $\dot{\mathbf{r}} = -9{\cdot}81 t\mathbf{j} + \mathbf{a}$

where $\mathbf{a}$ is a constant vector.

When $t = 0$, $\dot{\mathbf{r}}_{t=0} = 9{\cdot}81\sqrt{3}\mathbf{i}+9{\cdot}81\mathbf{j}$

hence $\mathbf{a} = 9{\cdot}81\sqrt{3}\mathbf{i}+9{\cdot}81\mathbf{j}$

Therefore $\dot{\mathbf{r}} = 9{\cdot}81\sqrt{3}\mathbf{i}+9{\cdot}81(1-t)\mathbf{j}$ . . . I

Again $\mathbf{r} = 9{\cdot}81\sqrt{3}t\mathbf{i}+9{\cdot}81(t-\frac{1}{2}t^2)\mathbf{j}+\mathbf{b}$

where **b** is the second constant vector.

When $t = 0$, however, $\mathbf{r} = \mathbf{0}$: therefore $\mathbf{b} = \mathbf{0}$.

The path of the projectile is given by

$$\mathbf{r} = 9{\cdot}81\sqrt{3}t\mathbf{i}+4{\cdot}905t(2-t)\mathbf{j} \qquad \ldots \qquad \ldots \text{II}$$

which is an equation of the second degree and describes a parabola.

(i) The time taken for the particle to reach the horizontal plane again is given by II when the vertical component, that is parallel to vector **j**, is zero. This occurs when $t = 2$ seconds.

(ii) The horizontal range at this time is given by the horizontal component of II, that is in the direction of vector **i**, when $t = 2$ seconds. Hence the range is $9{\cdot}81\sqrt{3}(2) = 19{\cdot}62\sqrt{3}$ metres.

(iii) A parabola is symmetrical about its axis. The greatest height, therefore, occurs when $t = 1$ second, and the vertical component of II is then

$$h = 4{\cdot}905(1) = 4{\cdot}905 \text{ metres}$$

Alternatively we can see from I that the vertical component of the velocity vanishes when $t = 1$ second, and we then substitute this value of t in II to obtain the greatest height 4·905 metres.

Summary

For a constant vector $\mathbf{r}$ we have

$$\frac{d\mathbf{r}}{dt} = \mathbf{0}$$

For a variable vector $\mathbf{r}$ where $\mathbf{r} = \mathbf{p}+\mathbf{q}$

then $$\frac{d\mathbf{r}}{dt} = \frac{d\mathbf{p}}{dt}+\frac{d\mathbf{q}}{dt}$$

For a constant coefficient a

$$\mathbf{r} = a\mathbf{p}$$

$$\frac{d\mathbf{r}}{dt} = a\frac{d\mathbf{p}}{dt}$$

For a parameter u

$$\mathbf{r} = u\mathbf{p}$$

$$\frac{d\mathbf{r}}{dt} = u\frac{d\mathbf{p}}{dt}+\frac{du}{dt}\mathbf{p}$$

For a dot product

$$\mathbf{r} = \mathbf{p}.\mathbf{q}$$

$$\frac{d\mathbf{r}}{dt} = \mathbf{p}.\frac{d\mathbf{q}}{dt}+\frac{d\mathbf{p}}{dt}.\mathbf{q}$$

For a cross product

$$\mathbf{r} = \mathbf{p}\times\mathbf{q}$$

$$\frac{d\mathbf{r}}{dt} = \mathbf{p}\times\frac{d\mathbf{q}}{dt}+\frac{d\mathbf{p}}{dt}\times\mathbf{q}$$

When s is the arc length of a curve, we have

$$\frac{d\mathbf{r}}{ds} = \hat{\mathbf{t}}$$

where $\hat{\mathbf{t}}$ is the unit tangent at a point on the curve.

$$\frac{d\mathbf{r}}{ds} = \frac{dx}{ds}\mathbf{i}+\frac{dy}{ds}\mathbf{j}+\frac{dz}{ds}\mathbf{k}$$

where $\frac{dx}{ds}$, $\frac{dy}{ds}$ and $\frac{dz}{ds}$ are the direction cosines of the unit tangent $\hat{\mathbf{t}}$.

Equation of tangent

$$\mathbf{T} = \mathbf{r} + u\hat{\mathbf{t}}$$

$$\frac{d\hat{\mathbf{t}}}{ds} = \frac{\hat{\mathbf{n}}}{\rho}$$

where $\hat{\mathbf{n}}$ is the unit principal normal and ρ the radius of curvature given by

$$\frac{1}{\rho} = \frac{d\psi}{ds}$$

where ψ is the angle made by the tangent with the positive direction of the x axis.

Equation of normal

$$\mathbf{N} = \mathbf{r} + u\hat{\mathbf{n}}$$

Radial component of velocity $= \dot{r}$

Transverse component of velocity $= r\dot{\theta}$

Radial component of acceleration $= \ddot{r} - r\dot{\theta}^2$

Transverse component of acceleration $= \frac{1}{r}\frac{d}{dt}(r^2\dot{\theta})$

Normal component of acceleration $= \frac{v^2}{\rho} = \dot{s}\dot{\psi}$

Tangential component of acceleration $= \dot{v} = \ddot{s}$

Integration: when $\frac{d\mathbf{F}}{dt} = \mathbf{r}$, then

$$\mathbf{F} = \int \mathbf{r}\,dt + \mathbf{c} \quad \text{or} \quad \int_{t_1}^{t_2} \mathbf{r}\,dt = \mathbf{F}(t_2) - \mathbf{F}(t_1)$$

Exercise 16

1 Find the velocity and acceleration vectors of the following displacement vectors

(i) $\sin t\,\mathbf{i}+\cos t\,\mathbf{j}+\sin 2t\,\mathbf{k}$

(ii) $3t\mathbf{i}+t^2\mathbf{j}-3t^3\mathbf{k}$

(iii) $(\sin t-\cos t)\mathbf{i}+(3\sin t+2\cos t)\mathbf{j}-\cos 2t\mathbf{k}$

(iv) $e^{2t}\mathbf{i}-e^t\mathbf{j}+2e^{3t}\mathbf{k}$

(v) $(t+1)\mathbf{i}-(t^2+t^3)\mathbf{j}+e^{-t}\mathbf{k}$

2 Let $\mathbf{r}_1 = t^2\mathbf{i}+t\mathbf{j}$ and $\mathbf{r}_2 = 3t\mathbf{i}-t^2\mathbf{j}$.

Find the first derivatives with respect to time of the following

(i) $\mathbf{r}_1+\mathbf{r}_2$ (ii) $\mathbf{r}_2-\mathbf{r}_1$ (iii) $\mathbf{r}_1.\mathbf{r}_2$ (iv) $\mathbf{r}_1\times\mathbf{r}_2$ (v) $\mathbf{r}_2\times\mathbf{r}_1$

3 Find the following integrals

(i) $\int(t\mathbf{i}+2t\mathbf{j}+\mathbf{k})dt$

(ii) $\int(\sin t\,\mathbf{i}+\cos t\,\mathbf{j})dt$

(iii) $\int(t^3\mathbf{i}+t^2\mathbf{j}+2t\mathbf{k}).(\mathbf{i}+\mathbf{j})dt$

4 (i) What is the unit vector $\hat{\mathbf{t}}$, with the direction of the tangent to a curve whose vector is $\mathbf{r}$, in terms of Cartesian coordinates?

(ii) Give the vector equation of the tangent line at a point P on a curve.

5 (i) What is the unit principal normal to a curve in terms of the tangent, $\mathbf{t}$, and the path, s?

(ii) Give the vector equation of the normal plane at the point P.

6 Find the vector equation of the tangent and normal to the circle $\mathbf{r} = \cos\phi\,\mathbf{i}+\sin\phi\,\mathbf{j}$ for $\phi = \pi/3$.

7 Find the vector equation of the tangent and normal to the circle $\mathbf{r} = (2+3\cos\phi)\mathbf{i}+(3\sin\phi-4)\mathbf{j}$ at the point $(5, -4)$.

8 What are the vector equations of the normal planes to the circle $\mathbf{r} = \cos\theta\,\mathbf{i}+\sin\theta\,\mathbf{j}$ when θ is (i) 0 radians and (ii) $\pi/4$?

9 What is the equation of the tangent and normal to the parabola $\mathbf{r} = ap^2\mathbf{i}+2ap\mathbf{j}$ at the point $P(ap_1^2, 2ap_1)$?

10 Let the path of a projectile be

$$\mathbf{r} = u(\cos\alpha)t\mathbf{i}+\{u(\sin\alpha)t-\tfrac{1}{2}gt^2\}\mathbf{j}$$

where u is the initial velocity, α is the angle of projection and t is the time measured in seconds from the time of projection. If $u = 29{\cdot}43$ m s^{-1}, $g = 9{\cdot}81$ m s^{-2} and $\alpha = 45°$, find

(i) the time for the projectile to reach its highest point

(ii) the maximum height of the projectile

(iii) the velocity of the projectile when it hits the horizontal plane through the point of projection

(iv) the velocity of the projectile at its highest point.

11 A point starts from (4, 0, 0) when $t = 0$ and is at $(4, 0, \pi)$ when $t = 1$. It rotates about the z axis at 5 revolutions second^{-1} and moves along the circular helix

$$\mathbf{r} = a\cos\theta\,\mathbf{i}+a\sin\theta\,\mathbf{j}+b\,\theta\,\mathbf{k}$$

Find the velocity and acceleration vectors of the point at the beginning of motion and when $t = 1$ second.

12 Find the displacement, velocity and acceleration vectors after time $t = 2$ of a particle moving along a curve such that

$$x = 4t^3-3t^2+1, \quad y = 3t^2-t, \quad z = t+1$$

13 What are the resolutes of the velocity and acceleration of the particle in question 12 in the direction of the vector $2\mathbf{i}-\mathbf{j}+2\mathbf{k}$ for $t = 2$?

14 A particle has an acceleration at time t given by

$$\ddot{\mathbf{r}} = \cos t\,\mathbf{i}+\sin t\,\mathbf{j}-g\,\mathbf{k}.$$

If it is projected from the origin with a velocity of 36 m s^{-1} along a line with direction ratios 4:7:4, find the velocity and position vectors of the particle at time t.

15 Find the radial and transverse components of the velocity and acceleration at time t of a particle moving along the curve $r = a\sin\theta$.

16 Find the radial and transverse components of the velocity and acceleration of a particle moving along the curves given by

(i) $r = a\theta$ (ii) $r = a/\theta$ for $\theta = \frac{1}{2}t$ in each case

(iii) $r = a\theta^2$ (iv) $r = a\cos\theta$ for $\theta = t$

17 Find the radial and transverse components of the velocity and acceleration of a particle which moves along the logarithmic spiral $r = e^{a\theta}$ when the angular velocity is constant.

18 Find the tangential and normal components of the acceleration of the particle moving along the curve $\mathbf{r} = \cos t\,\mathbf{i}+\sin t\,\mathbf{j}+t^2\,\mathbf{k}$.

19 Show that the acceleration of a particle moving along the circle $\mathbf{r} = r\cos\theta\,\mathbf{i}+r\sin\theta\,\mathbf{j}$ with constant speed is constant in magnitude and direction towards the centre of the circle.

20 The velocities of a particle along and perpendicular to a radius vector from a fixed origin are $\lambda^2 r$ and $\mu\theta^2$. Find the polar equation of the path of the particle and also the component accelerations in terms of r and θ. (*CS*)

5 · *Miscellaneous examples*

In the following examples, unit vectors along the Ox and Oy axes are represented by $\mathbf{i}$ and $\mathbf{j}$ respectively.

Example 1

Two ships, A and B, have the following displacement and velocity vectors at 8 a.m.

$$\mathbf{a} = 8\mathbf{i}-5\mathbf{j} \quad \text{and} \quad \mathbf{v}_A = -3\mathbf{i}+7\mathbf{j}$$

$$\mathbf{b} = -5\mathbf{i}+\mathbf{j} \quad \text{and} \quad \mathbf{v}_B = \mathbf{i}-2\mathbf{j}$$

where the units are nautical miles and knots.

What are their respective position vectors at 10 a.m.?

Find the distance between the ships at 10 a.m.

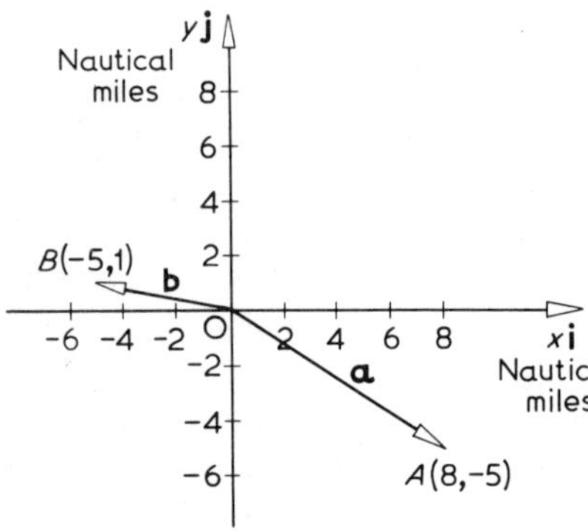

Figure 5.1

Figure 5.2

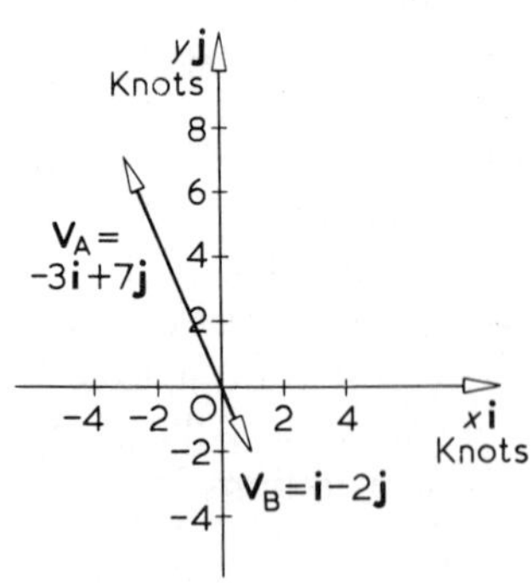

Figure 5.1 shows the position vectors of the two ships at 8 a.m. A moves along a course parallel to the vector $\mathbf{v}_A$, shown in Figure 5.2.

Since the time taken is 2 hours, the ship moves through a displacement represented by $2(-3\mathbf{i}+7\mathbf{j}) = -6\mathbf{i}+14\mathbf{j}$ from A (Figure 5.3).

Figure 5.3

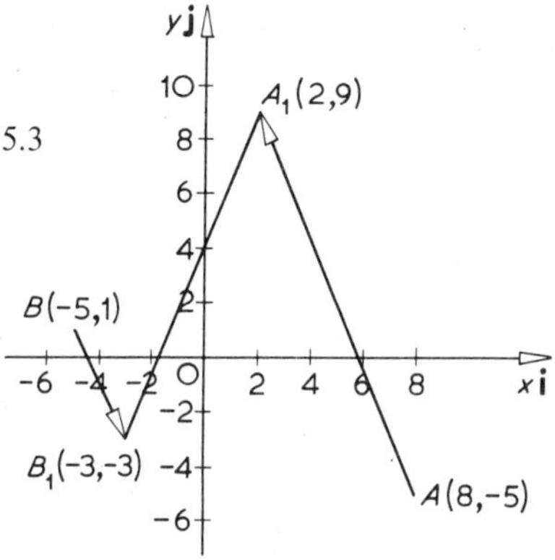

Therefore A arrives at A_1 given by the vector

$$\mathbf{OA}_1 = (-6\mathbf{i}+14\mathbf{j})+(8\mathbf{i}-5\mathbf{j})$$
$$= 2\mathbf{i}+9\mathbf{j}$$

Similarly the course followed by B during the same two hours is shown on Figure 5.3 and its position at $\mathbf{B}_1$ is given by

$$\mathbf{OB}_1 = 2(\mathbf{i}-2\mathbf{j})+(-5\mathbf{i}+\mathbf{j})$$
$$= -3\mathbf{i}-3\mathbf{j}$$

The respective position vectors of the ships, therefore, are

$$A = 2\mathbf{i}+9\mathbf{j} \qquad B = -3\mathbf{i}-3\mathbf{j}$$

The distance between the ships at 10 a.m. is given by the magnitude of the vector $\mathbf{B}_1\mathbf{A}_1$, that is

$$|2\mathbf{i}+9\mathbf{j}-(-3\mathbf{i}-3\mathbf{j})| = |5\mathbf{i}+12\mathbf{j}|$$
$$= \sqrt{5^2+12^2}$$

Therefore their distance apart is 13 nautical miles.

Example 2

Two spheres, A of mass 3 kg and B of mass 4 kg, move with velocity vectors

$$\mathbf{u}_A = 4\sqrt{3}\mathbf{i}-4\mathbf{j} \quad \text{and} \quad \mathbf{u}_B = -4\mathbf{i}-3\sqrt{3}\mathbf{j}$$

The equation of their line of centres on impact is $\mathbf{r} = t\mathbf{i}$.

If the coefficient of restitution e is $\frac{3}{4}$, find the velocity vectors after impact.

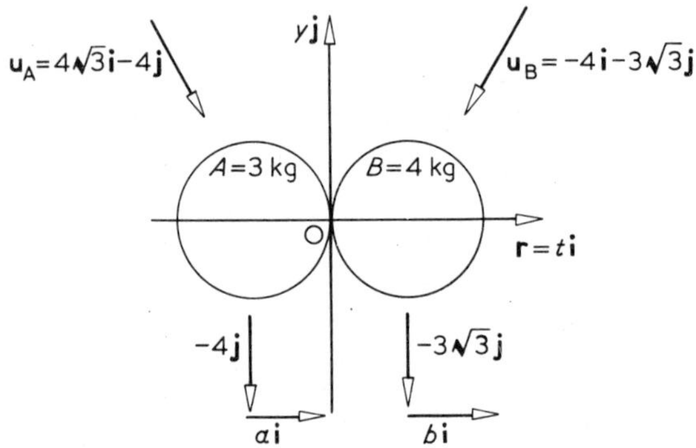

Figure 5.4

The components of the velocities perpendicular to the line of centres is unaltered by the impact. Therefore A retains its $-4\mathbf{j}$ velocity component and B retains its $-3\sqrt{3}\mathbf{j}$ velocity component.

Let the velocity components parallel to the line of centres be $a\mathbf{i}$ for A and $b\mathbf{i}$ for B.

By the principle of the conservation of momentum, we have

$$\text{momentum before impact} = \text{momentum after impact}$$

$$3(4\sqrt{3})+4(-4) = 3a+4b \quad \ldots \quad \ldots \text{I}$$

By Newton's law of restitution, we have

$$e(\text{velocity of approach}) = \text{velocity of separation}$$

$$\tfrac{3}{4}(4\sqrt{3}+4) = -a+b \quad \ldots \quad \ldots \text{II}$$

From I and II, we have $a = -4$ and $b = 3\sqrt{3}-1$ so that the velocities after impact are

$$\mathbf{v}_A = -4\mathbf{i}-4\mathbf{j} \quad \text{and} \quad \mathbf{v}_B = (3\sqrt{3}-1)\mathbf{i}-3\sqrt{3}\mathbf{j}$$

Example 3

The point A is fixed and $\mathbf{OA} = 4\mathbf{i}+6\mathbf{j}$; the point P is variable and $\mathbf{OP} = 2\mathbf{i}\cos\theta+2\mathbf{j}\sin\theta$, where θ varies. If $\mathbf{OA}+\mathbf{OP} = 2\mathbf{OQ}$, show that the locus of Q is a circle and find its centre and radius. (*C*)

$$\mathbf{OA}+\mathbf{OP} = 4\mathbf{i}+6\mathbf{j}+2\mathbf{i}\cos\theta+2\mathbf{j}\sin\theta$$

$$= (4+2\cos\theta)\mathbf{i}+(6+2\sin\theta)\mathbf{j}$$

Therefore $\mathbf{OQ} = (2+\cos\theta)\mathbf{i}+(3+\sin\theta)\mathbf{j}$

This vector represents the locus of a point moving in one plane on the circumference of a circle centre (2, 3) since the distance of any point on the circumference from (2, 3) is

$$\{2-(2+\cos\theta)\}^2+\{3-(3+\sin\theta)\}^2 = \cos^2\theta+\sin^2\theta = 1,$$

a constant.

Therefore the centre of the circle is (2, 3) and its radius is 1 unit.

Example 4

A point A of a lamina moves with constant acceleration f along a line Ox fixed in the plane of the lamina, and the lamina also rotates about an axis through A, perpendicular to its plane, with constant angular velocity ω. Prove that, at any instant of time, points of the lamina whose accelerations have magnitude $a\omega^2$ lie on a circle which is stationary relative to A, and find the radius of this circle and its centre relative to A.

If initially A had velocity u along Ox, find, relative to A, the locus of points of the lamina whose speeds are $b\omega$ at time t. (*L*)

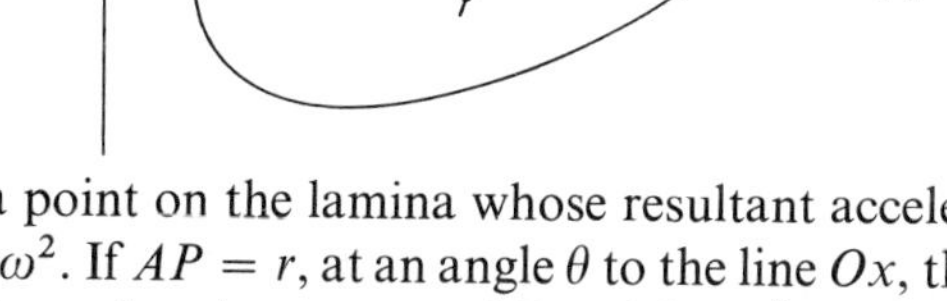

Figure 5.5
acceleration

Let P be a point on the lamina whose resultant acceleration is of magnitude $a\omega^2$. If $AP = r$, at an angle θ to the line Ox, then we have the resultant acceleration vector of P as $(f-r\omega^2\cos\theta)\mathbf{i}+r\omega^2\sin\theta\,\mathbf{j}$, that is $(f-\omega^2X)\mathbf{i}+\omega^2Y\mathbf{j}$ where the coordinates of P are (X, Y).

Equating the square of the magnitude of this vector to the square of the resultant magnitude, we have

$$(f-\omega^2 X)^2+(\omega^2 Y)^2 = a^2\omega^4$$

that is $\quad \omega^4 X^2+\omega^4 Y^2-2f\omega^2 X+f^2-a^2\omega^4 = 0.$

Hence, we have the locus of points which satisfy the equation

$$x^2+y^2-\frac{2f}{\omega^2}x+\frac{f^2-a^2\omega^4}{\omega^4} = 0$$

and this represents the equation of a circle, centre $\left(\frac{f}{\omega^2}, 0\right)$, and radius $\sqrt{\frac{f^2}{\omega^4}-\frac{f^2}{\omega^4}+a^2} = a$. Since f, ω and a are all constants, we have, therefore, a circle which is stationary relative to A.

Figure 5.6
velocity

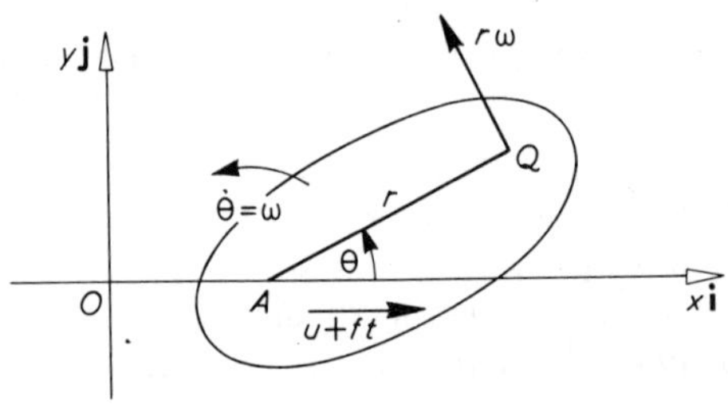

Let Q be a point on the lamina whose resultant velocity is of magnitude b at time t. The resultant velocity vector of Q is

$$(u+ft-r\omega\sin\theta)\mathbf{i}+r\omega\cos\theta\,\mathbf{j}$$

that is $\quad (u+ft-\omega Y)\mathbf{i}+\omega X\mathbf{j}$

where the coordinates of Q are (X, Y).

Again equating the square of the magnitude of this vector to the square of the resultant speed, we have

$$(u+ft)^2+\omega^2 Y^2-2(u+ft)\omega Y+\omega^2 X^2 = b^2\omega^2$$

that is we have the locus of points satisfying the equation

$$x^2+y^2-\frac{2(u+ft)}{\omega}y+\left(\frac{u+ft}{\omega}\right)^2-b^2 = 0.$$

This again is a circle, but its centre is $\left(0, \frac{u+ft}{\omega}\right)$ and so its ordinate, relative to A, varies with time. The radius is b.

Example 5

Three forces are represented completely by the vectors $p\mathbf{AD}$, $q\mathbf{BE}$, $r\mathbf{CF}$, where D, E, F are points on the sides BC, CA, AB, respectively, of the triangle ABC such that

$$\frac{BD}{DC} = l, \qquad \frac{CE}{EA} = m, \qquad \frac{AF}{FB} = n$$

Show that the three forces are equivalent to three forces acting along the sides of the triangle and find these forces.

If $p = q = r$ and $l = m = n$, show that the system reduces to a couple and find its magnitude in terms of p, l and the area of the triangle. *(C)*

Figure 5.7

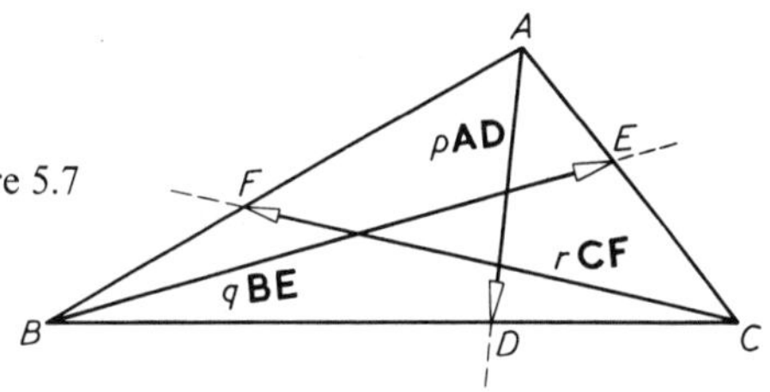

Using the notation of the important example on page 22, we have

$$p\mathbf{AD} = \lambda\mathbf{AB} + \mu\mathbf{AC} \quad \text{where} \quad \lambda + \mu = p \quad \text{and} \quad \frac{\mu}{\lambda} = l$$

Therefore $\lambda(1+l) = p$, whence

$$\lambda = \frac{p}{1+l} \quad \text{and} \quad \mu = \frac{pl}{1+l}$$

Hence
$$p\mathbf{AD} = \frac{p}{1+l}\mathbf{AB} + \frac{pl}{1+l}\mathbf{AC}$$

$$q\mathbf{BE} = \frac{q}{1+m}\mathbf{BC} + \frac{qm}{1+m}\mathbf{BA}$$

$$r\mathbf{CF} = \frac{r}{1+n}\mathbf{CA} + \frac{rn}{1+n}\mathbf{CB}$$

The three forces on the left-hand side of these equations are together equivalent to a force

$$\left(\frac{p}{1+l} - \frac{mq}{1+m}\right)\mathbf{AB} \qquad \text{along } AB$$

$$\left(\frac{q}{1+m}-\frac{nr}{1+n}\right)\mathbf{BC} \quad \text{along } BC$$

$$\left(\frac{r}{1+n}-\frac{pl}{1+l}\right)\mathbf{CA} \quad \text{along } CA.$$

When $p = q = r$ and $l = m = n$, we have

$$\left(\frac{p}{1+l}-\frac{pl}{1+l}\right)\mathbf{AB}+\left(\frac{p}{1+l}-\frac{pl}{1+l}\right)\mathbf{CA}+\left(\frac{p}{1+l}-\frac{pl}{1+l}\right)\mathbf{BC}$$

We can rewrite this system of forces as

$$\frac{p}{1+l}\left(\mathbf{AB}+\mathbf{BC}\right)-\frac{p}{1+l}\mathbf{AC}+\frac{pl}{1+l}\left(\mathbf{AC}+\mathbf{BA}\right)-\frac{pl}{1+l}\mathbf{BC}$$

Since $\mathbf{AB}+\mathbf{BC} = \mathbf{AC}$, the first two terms represent forces which are equal and opposite, one acts through B and the other acts in **CA**. Similarly, the next two terms represent equal and opposite forces, one acts through A and the other in **BC**.

The system, therefore, reduces to two couples which are together equal to a couple.

We find the magnitude of this couple by taking moments about some convenient point. Let us take moments about A for the three forces acting in the sides of the triangle

$$\frac{p}{1+l}BC.h-\frac{pl}{1+l}BC.h$$

where h is the altitude of the triangle. The two forces acting through A have, of course, no moment about A.

Therefore the magnitude of the couple is

$$2\Delta p\frac{1-l}{1+l}$$

where Δ represents the area of the triangle.

Example 6

ABC is an isosceles right-angled triangle such that $AB = 2a$ and angle ACB is a right angle. A particle is moved in the plane ABC under the action of a force of repulsion of magnitude kAP^2 along AP.

Prove, by direct integration in each case, that the total work done by the repelling force has the same value $8a^3k/3$ when the particle is moved from A to B (i) along the straight line AB, (ii) the semicircle ACB, (iii) the path formed by the straight line AC and the quadrant arc CB. *(C)*

Figure 5.8

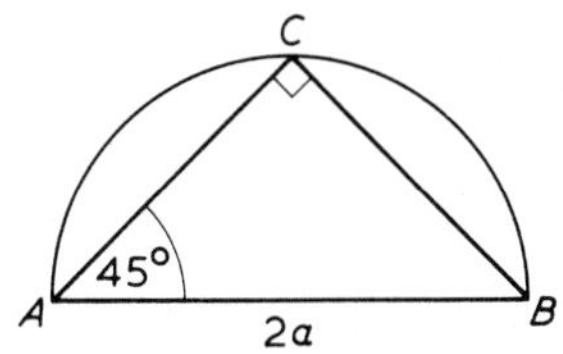

Let **r** be the position vector of the particle P with respect to A.
(i) When the particle moves along the diameter AB, we have

$$\text{Work done} = k\int_0^{2a} \mathbf{r}^2 . d\mathbf{r}$$

$$= k\left[\frac{r^3}{3}\right]_0^{2a} = \frac{8a^3k}{3}$$

(ii) Along the semicircle, we have $r = 2a\cos\theta$ and $dr = -2a\sin\theta\, d\theta$

$$\text{Work done} = k\int_0^{2a} \mathbf{r}^2 . d\mathbf{r}$$

that is

$$\text{Work done} = 4a^2k\int_{\pi/2}^{0} \cos^2\theta\,(-2a\sin\theta)\, d\theta$$

$$= \left[\frac{8a^3k}{3}\cos^3\theta\right]_{\pi/2}^{0}$$

$$= \frac{8a^3k}{3}$$

(iii) Consider the work done by the force alone **AC**.

$$\text{Work done} = k\int_0^{a\sqrt{2}} \mathbf{r}^2 . d\mathbf{r}$$

$$= k\left[\frac{r^3}{3}\right]_0^{a\sqrt{2}} = 2\sqrt{2}\, a^3k/3$$

Next we consider the work done along the quadrant arc CB.

$$\text{Work done} = 4a^2k\int_{\pi/4}^{0} \cos\theta(-2a\sin\theta)d\theta$$

$$= \frac{8a^3k}{3}\left[\cos^3\theta\right]_{\pi/4}^{0}$$

$$= 8a^3k/3 - 2\sqrt{2}\,a^3k/3$$

Therefore the total work done by the force on the particle is

$$2\sqrt{2}\,a^3k/3 + 8a^3k/3 - 2\sqrt{2}\,a^3k/3$$

$$= 8a^3k/3$$

Therefore the total work done in each case is $8a^3k/3$.

Example 7

A particle moves in a plane in such a way that its tangential and normal components of acceleration are always constant. Prove that in general the particle describes a logarithmic spiral $s = Ae^{k\psi}$, where A and k are constants. *(O & C)*

The acceleration along the tangent is

$$\ddot{s} = a, \text{ a constant.}$$

Hence $\dot{s} = at + b$, where b is the constant of integration.

Again $s = \frac{1}{2}at^2 + bt + c$, the second constant of integration. . . I

The acceleration along the normal is

$$\dot{s}\dot{\psi} = l, \text{ a constant}$$

that is $\dot{\psi} = \dfrac{l}{\dot{s}} = \dfrac{l}{at+b}$

Hence $\psi = \dfrac{l}{a}\log_e(at+b) + m$,

where m is the constant of integration.

$$= N\log_e(at+b),$$

where the constants have been combined in N.

Therefore $at + b = e^{\psi/N}$.

From I, we have

$$s = A(at+b)^2 = Ae^{k\psi} \quad \text{for constants } A \text{ and } k$$

Revision exercises

I

1 Let **a**, **b** and **c** have components [2, −1, 0], [1, 0, −1] and [0, 2, 1], respectively.
Find the components of the following vectors
(i) $2\mathbf{a}+3\mathbf{b}$ (ii) $4\mathbf{b}-\mathbf{c}$ (iii) $\mathbf{a}+\mathbf{b}+\mathbf{c}$ (iv) $\mathbf{a}-\mathbf{b}-\mathbf{c}$
(v) $2\mathbf{a}-\mathbf{b}+3\mathbf{c}$

2 Find the equations of the lines which pass through the following points and have the corresponding direction ratios
(i) $P(1, 0, 3)$ and $1:-1:2$ (ii) $Q(-1, 2, 1)$ and $3:2:-2$
(iii) $R(3, 4, 5)$ and $2:-1:5$

3 Two forces $\mathbf{F}_1 = 2\mathbf{i}+3\mathbf{j}+6\mathbf{k}$ and $\mathbf{F}_2 = 2\mathbf{i}+2a\mathbf{j}+a\mathbf{k}$ act through points whose vectors are $\mathbf{S}_1 = \mathbf{i}-2\mathbf{j}+2\mathbf{k}$ and $\mathbf{S}_2 = \mathbf{i}-6\mathbf{k}$ respectively, relative to a fixed point and three mutually perpendicular unit vectors **i**, **j** and **k**.
Show that the two forces can never be equal and find the value of a if the lines of action of $\mathbf{F}_1$ and $\mathbf{F}_2$ intersect.
Find also the vector equation of the line of action of the resultant of $\mathbf{F}_1$ and $\mathbf{F}_2$.

4 A ball is moving with velocity vector $\mathbf{v} = 12\sqrt{2}\mathbf{i}-12\sqrt{2}\mathbf{j}$ and impinges on the smooth fixed coordinate plane $y = 0$. If the coefficient of restitution is $\frac{3}{4}$ find the velocity vector of the ball after impact.

5 The position vector of a moving point is given by
$$\mathbf{r} = (1+5\sin\theta)\mathbf{i}+(5+3\cos\theta)\mathbf{j}+(6+4\cos\theta)\mathbf{k}$$
where the units are in metres. Show that the locus described by the point is a circle and find its centre and radius. Find also the inclination of its line of greatest slope to the plane $z = 0$.
If a particle is projected in a straight line from the origin to the nearest point of the circle to the plane $z = 0$, find its velocity vector if the time taken by the particle is $\frac{1}{3}$ second and the velocity is constant.

II

1 Show that vectors **a**, **b** and 2**a** − **b** are collinear.

2 If O is the origin, write down the equation of a plane defined by three given points which are contained by the plane and whose vectors are **a**, **b** and **c**, relative to the origin.

3 Find the work done by constant forces with magnitudes 3, 2 and 1 newtons in the directions of vectors [6, 2, 9], [2, −9, 6] and [9, −2, −6] respectively, acting on a particle which is moved from $A(2, 1, -1)$ to $B(4, 7, 1)$, where the units are metres.

4 A ball moving with velocity vector $\mathbf{v} = 2\sqrt{3}\mathbf{i} - 2\mathbf{j}$ strikes the smooth plane $y = 0$. If the coefficient of restitution is $\frac{1}{2}$ find the velocity vector after impact.

5 On a vast expanse of flat sand sit three turtles: Fred, Charlie and Liz. Their position vectors are

Fred: $-5\mathbf{i}+5\mathbf{j}$ Charlie: $-\mathbf{i}-5\mathbf{j}$ Liz: $3\mathbf{i}+\mathbf{j}$

If Liz sits still whilst Fred and Charlie simultaneousely move towards her at their maximum speed of $\sqrt{10}$ m s^{-1}, find how long it is for the first turtle to reach her.

If Charlie were not there and Liz moves away with velocity vector **i** at the same time as Fred moves, find Fred's least time to reach Liz and his position vector.

III

1 Show that the points (1, 2, 4), (2, −1, 0) and (3, −4, −4) are collinear.

2 Find the equations of the planes containing the following sets of three points

(i) $A(1, 2, -1), B(2, 1, 0), C(0, 1, -2)$

(ii) $P(2, 1, 0), Q(-1, 3, 1), R(1, 0, -1)$.

3 A ball with velocity vector $\mathbf{v} = 6\mathbf{i} - 6\sqrt{3}\mathbf{j}$ hits the fixed elastic plane $y = 0$. If the impact turns the direction of motion through a right angle, find the coefficient of restitution.

4 Find the vector of the point of intersection of the parabola

$\mathbf{r}_1 = 2p\mathbf{i}+p^2\mathbf{k}$ and the helix $\mathbf{r}_2 = \cos\theta\,\mathbf{i}+\sin\theta\,\mathbf{j}+\dfrac{\theta}{4\pi}\mathbf{k}$

This vector now describes a circle whose centre is given by vector $\frac{1}{4}\mathbf{k}$. Write down the equation of this circle and when

$\theta = 2n\pi$, for successive integral values of n, show that the distances between this circle and the helix increase by one-half a unit.

5 Two projectiles have velocity vectors $\mathbf{v}_1 = a\mathbf{i}+(b-gt)\mathbf{j}$ and $\mathbf{v}_2 = c\mathbf{i}+(d-gt)\mathbf{j}$.
Show that their relative displacement is proportional to the time of flight and constant in direction. Show also that their relative velocity is constant. In particular when $b+c = a+d$, prove that one projectile in flight is always at an angle of 45° to the other and when $a = c$, one projectile is directly below the other.

IV

1 Find a vector that is parallel to the line joining $P(3, -4, 5)$ and $Q(5, -5, 8)$.

2 What is the equation of the plane parallel to the plane $2x-3y+2z-1 = 0$ and which passes through the point $(1, -1, 2)$? What is the distance between these two planes?

3 A sphere with velocity vector $\mathbf{v} = 10\sqrt{3}\mathbf{i}+10\mathbf{j}$ strikes an identical sphere which is at rest. The equation of the line of centres at that instant is $\mathbf{r} = t\mathbf{j}$. If $e = \frac{1}{2}$, find the velocity vectors of the spheres after impact.

4 Find the pitch of the helix $\mathbf{r} = 2\cos\theta\,\mathbf{j}+2\sin\theta\,\mathbf{k}+\dfrac{6\theta}{\pi}\mathbf{i}$. Determine vectors $\mathbf{r}_1$ and $\mathbf{r}_2$ of the points of intersection of the helix with the line $\mathbf{l} = \mathbf{j}+\sqrt{3}\mathbf{k}+2\mathbf{i}+2t(\mathbf{i}-\mathbf{j})$ and show that $|\mathbf{r}_1-\mathbf{r}_2| = 2\sqrt{2}$.

5 A fly takes off when $t = 0$ along a parabolic path given by $\mathbf{r} = t^2\mathbf{i}+2t\mathbf{j}$ where t is measured in seconds. A light source is at the point $50\mathbf{i}+40\mathbf{j}+10\mathbf{k}$. Show that when $t = 5$, its shadow crosses the line which goes through the point $40\mathbf{i}+50\mathbf{j}+20\mathbf{k}$ and parallel to $6\mathbf{i}+5\mathbf{j}+\mathbf{k}$. What is the vector of this crossing point?
The fly now flies to this crossing point where there is a conveniently situated monorail to the origin. It slides along it at a speed of $\sqrt{209}$ cm s^{-1}. Find the time taken for the slide. All the distances are in centimetres.

V

1 Show that vectors $2\mathbf{a}+\mathbf{b}$, $3\mathbf{a}-2\mathbf{b}$ and $4\mathbf{a}-5\mathbf{b}$ are collinear.

2 Sketch the following curves

(i) $\mathbf{r}_1 = a\cos p\,\mathbf{i}+b\sin p\,\mathbf{j}+cp\mathbf{k}$

(ii) $\mathbf{r}_2 = a\cos p\,\mathbf{i}+a\sin p\,\mathbf{j}+b\mathbf{k}$

3 A ball with a velocity vector $\mathbf{v} = 2\mathbf{i}-2\sqrt{3}\mathbf{j}$ strikes a smooth plane $x = 0$. If the coefficient of restitution is $\frac{1}{2}$, find the velocity vector after impact.

4 Rays from the origin generate the ellipse

$$\mathbf{r} = 5\cos\phi\,\mathbf{i}+b\sin\phi\,\mathbf{j}+\mathbf{k}$$

where ϕ is a parameter. If the eccentricity of the ellipse is 4/5, find the value of b.

What is the equation of the ellipse formed by these rays in the plane $z = 2$?

Write down the general form of the equations of ellipses formed in the planes $z = n$, for $n > 0$.

5 Find the resultant $\mathbf{F}$ of two forces $\mathbf{A}$ and $\mathbf{B}$ where $\mathbf{A}$ is 9 newtons and has direction ratios 1:4:8 and $\mathbf{B}$ is $\sqrt{24}$ newtons and direction ratios $1:-1:-2$.

A particle free to move only in the plane $y = 0$ moves under the influence of $\mathbf{F}$ a distance of 6 metres. Find the work done on the particle.

VI

1 To what coordinate plane is the line joining the points $(3, -2, -1)$ and $(2, -2, 1)$ parallel?

2 A sphere of mass 4 kg with velocity vector $\mathbf{v} = 10\sqrt{3}\mathbf{i}+10\mathbf{j}$ strikes a second sphere of the same radius but of mass 6 kg which is at rest. The equation of the line of centres at that instant is $\mathbf{r} = t\mathbf{j}$. If $e = \frac{1}{2}$, calculate the velocity vectors after impact.

3 Show that the relative velocity of two particles moving under the acceleration of gravity is constant.

4 A marker buoy has vector $2\mathbf{i}-8\mathbf{j}$. At time $t = 0$, the buoy emits a dye which spreads outwards uniformly at a speed of $\frac{5}{8}$ knots. The dye is in a current moving with velocity $3\mathbf{i}+4\mathbf{j}$. Find the locus of the perimeter of the dye-affected area in terms of time and a suitable parameter.

A boat is at $17\mathbf{i}+12\mathbf{j}$ at $t = 0$ and moves directly against the flow of the current at an actual speed of 10 knots. Calculate the time and the position vector when the boat enters the dye-affected area. (Distances are in nautical miles and speeds in knots.)

5 For the circular helix $\mathbf{r} = a\cos\theta\,\mathbf{i}+a\sin\theta\,\mathbf{j}+b\theta\,\mathbf{k}$ a point starts at (4, 0, 0) when $t = 0$ and is at (4, 0, π) when $t = 1$. The point moves round the z axis at 5 revolutions second^{-1}. Find the position vector after time t. What is the rate of increase of the position vector for 1 second intervals?

VII

1 What are the conditions for vector $\mathbf{a} = [a_1, a_2, a_3]$ to be parallel to vector $\mathbf{b} = [b_1, b_2, b_3]$?
Illustrate these conditions by determining whether the points $A(2, 3, 4)$, $B(-3, -1, 1)$ and $C(-8, -5, 7)$ are collinear.

2 Find the velocity and acceleration vectors of a particle which moves so that at time t its position vector is given by

$$\mathbf{r} = a\cos t\,\mathbf{i}+a\sin t\,\mathbf{j}+bt\,\mathbf{k}$$

3 The point A is fixed and $\mathbf{OA} = 6\mathbf{i}+8\mathbf{j}+12\mathbf{k}$. The point P is variable and $\mathbf{OP} = 4\cos\theta\,\mathbf{i}+4\sin\theta\,\mathbf{j}+6\mathbf{k}$ where θ varies. If $\mathbf{OA}+\mathbf{OP} = 2\mathbf{OQ}$ show that the locus of Q is a circle and find its centre and radius.

4 Two forces $\mathbf{F}_1 = 2\mathbf{i}-8\mathbf{j}+6\mathbf{k}$ and $\mathbf{F}_2 = 2\mathbf{i}+2\mathbf{j}+6\mathbf{k}$ act through points whose vectors are $\mathbf{S}_1 = \mathbf{ai}+\mathbf{bj}-14\mathbf{k}$ and $\mathbf{S}_2 = 3\mathbf{i}+p\mathbf{j}+q\mathbf{k}$. Find the values of the constants if the lines of action of the forces meet at the point given by $\mathbf{S}_3 = -3\mathbf{i}-7\mathbf{j}-8\mathbf{k}$.
Show that the points whose vectors are $\mathbf{S}_4 = \mathbf{i}-13\mathbf{j}+4\mathbf{k}$ and $\mathbf{S}_5 = 5\mathbf{i}-19\mathbf{j}+16\mathbf{k}$ lie on the line of action of the resultant and find the work done on a particle which is moved from $\mathbf{S}_4$ to $\mathbf{S}_5$. Let the units be kg and metres.

5 Two spheres, A of mass 12 kg and B of mass 9 kg, move with velocity vectors

$$\mathbf{v}_A = 4\mathbf{i}-4\mathbf{j} \quad \text{and} \quad \mathbf{v}_B = -3\mathbf{i}-3\mathbf{j}$$

and the equation of the line of centres on impact is $\mathbf{r} = t\mathbf{i}$. If $e = \frac{2}{3}$, find the velocity vectors after impact.

VIII

1 Find the vector of the mid-point of $\mathbf{a} = 3\mathbf{i}+3\mathbf{j}-\mathbf{k}$ and $\mathbf{b} = \mathbf{i}-\mathbf{j}+3\mathbf{k}$.

2 Find the angular velocity about an axis of vector $\boldsymbol{\omega}$, modulus 3 radians second^{-1} and whose direction cosines are $\frac{2}{3}$, 1, $-\frac{1}{3}$.

3 What is the value of λ for which the three points $A(3, -1, 2)$, $B(5, \lambda, 0)$ and $C(4, 1, 1)$ are collinear?
If AC then rotates about C in the plane of the triangle OAB, find the minimum length of AB.

4 The velocity vector of a projectile is $\mathbf{v} = a\mathbf{i}+(b-gt)\mathbf{j}$. Find the time taken to go through the point whose position vector is $\mathbf{r} = c\mathbf{i}+d\mathbf{j}$ and show that $2a^2d+c^2g = 2abc$.
When $c = \sqrt{3}d$, show that the angle of the projection must be more than 30° to the horizontal.

5 A vessel is sighted in a position given by $\mathbf{s}_1 = 6\mathbf{i}-\mathbf{j}$ and its velocity is $\mathbf{v} = -20\mathbf{i}+10\mathbf{j}$ where the units are in nautical miles and knots. A launch capable of a speed of 10 knots wishes to intercept the vessel. The position of the launch is $\mathbf{s}_2 = -6\mathbf{i}+11\mathbf{j}$. What is the quickest time the launch takes to reach the vessel and the course it must set to do so?

IX

1 Show that the points $A(3, 1, -1)$, $B(7, -5, -5)$ and $C(5, -2, -3)$ are collinear and find the ratio in which B divides the line segment AC. Does your result show an internal or external division of AC?

2 For what positive values of a and t does the straight line $\mathbf{r}_1 = a(\mathbf{i}+\mathbf{j})+t\mathbf{k}$ intersect the helix $\mathbf{r}_2 = 2\cos p\,\mathbf{i}+2\sin p\,\mathbf{j}+p\mathbf{k}$? What is the length of each straight line intercept?

3 Find the angular velocity vector $\boldsymbol{\omega}$ of a rigid body which is rotating about an axis OR at an angular velocity of 9 radians second^{-1} and where R is the point $(1, 2, -2)$ relative to O.

4 The vectors of two points are given by

$$\mathbf{r}_1 = (1+5\sin\theta)\mathbf{i}+(5+3\cos\theta)\mathbf{j}+(6+4\cos\theta)\mathbf{k}$$
$$\mathbf{r}_2 = (1+3\cos\theta)\mathbf{i}+(5+4\cos\theta)\mathbf{j}+(6+5\sin\theta)\mathbf{k}$$

where θ is a parameter. Show that the locus of each point is a circle with the same centre and the same radius but with different inclinations to the plane $z = 0$. Find these inclinations and the line of intersection of the two planes formed by the circles.

5 Two identical smooth spheres A and B moving on a horizontal plane with velocity vectors $3v\mathbf{i}$ and $-v\mathbf{j}$ respectively, collide. The equation then of their line of centres is $\mathbf{r} = \lambda \cos\theta\,\mathbf{i} + \lambda \sin\theta\,\mathbf{j}$ where λ is a parameter. If the velocity vector of B after the collision is $u\mathbf{i}$ and the coefficient of restitution is 3/7, prove that $\tan\theta$ is either $\frac{1}{2}$ or 7 and in each case find u and the velocity vector of A after the collision.

X

1 O is the centre of a circle and its radius vector, represented by the unit vector $\hat{\mathbf{r}}$, rotates with angular speed ω, not necessarily constant. Let $\hat{\mathbf{t}}$ be the unit vector in the direction of the tangent at P, at right angles to $\hat{\mathbf{r}}$. Give expressions for $d\hat{\mathbf{r}}/dt$ and $d\hat{\mathbf{t}}/dt$.

2 Find the velocity and acceleration vectors of a particle moving along the curve $\mathbf{r} = r\cos\phi\,\mathbf{i} + r\sin\phi\,\mathbf{j}$.

3 A particle is projected from a plane whose line of greatest slope is given by $\mathbf{r} = p(\mathbf{i}+\mathbf{j})$. The flight of the projectile is in the vertical plane containing the line of greatest slope. If the initial velocity of the projectile is $\mathbf{v}_0 = a\mathbf{i} + b\mathbf{j}$, find the value of p when the projectile hits the plane again.
If the projectile hits the plane normally, find the angle of projection.

4 Two ships A and B have the following displacement and velocity vectors at 8 a.m.

$$\mathbf{s}_A = \mathbf{i} - 5\mathbf{j} \quad \text{and} \quad \mathbf{v}_A = 4\mathbf{i} + 6\mathbf{j}$$
$$\mathbf{s}_B = -3\mathbf{i} + \mathbf{j} \qquad\qquad \mathbf{v}_B = 3\mathbf{i} - \mathbf{j}$$

The units are nautical miles and knots. What are their respective position vectors at 10 a.m. and the distance between them at this time?

5 The vector of a point is given by

$$\mathbf{r} = (2+5\sin\theta)\mathbf{i} + (3+4\cos\theta)\mathbf{j} + (4+3\cos\theta)\mathbf{k}$$

Show that the figure described by the point is a circle and find its centre and radius.
Obtain the maximum and minimum distances from the plane $z = 0$ and the inclination of the plane of the circle to this plane.

XI

1 $\hat{\mathbf{u}}$ is a unit vector which rotates in a plane with angular velocity ω. In what direction and of what magnitude is the vector $d\hat{\mathbf{u}}/dt$?

2 O is the centre of a circle and the radius vector, $\mathbf{r}$, rotates with angular speed ω, not necessarily constant. Let $\mathbf{r} = r\hat{\mathbf{r}}$ and then find $\mathbf{v}$ in terms of r, ω and $\hat{\mathbf{s}}$, where s is the arc length. Find also the acceleration of the radius vector.

3 Let the magnitude of $\mathbf{r}$ be a variable scalar, r. Find $\dot{\mathbf{r}}$ and $\ddot{\mathbf{r}}$.

4 Let $\mathbf{v} = v\hat{\mathbf{t}}$ and show that $\dfrac{d}{dt}\left(v\hat{\mathbf{t}}\right) = \dfrac{d}{dt}\hat{\mathbf{t}} + kv^2\hat{\mathbf{n}}$, where k is the curvature and $\hat{\mathbf{n}}$ the unit principal normal.

5 Find the normal and tangential components of acceleration of the curve

$$\mathbf{r} = t\mathbf{i} + t^2\mathbf{j} + \tfrac{1}{3}t^3\mathbf{k}$$

at $t = 0$ and at $t = 3$.

XII

1 Find the unit tangent $\hat{\mathbf{t}}$ and the unit principal normal $\hat{\mathbf{n}}$ of the circular helix

$$\mathbf{r} = [a\cos p,\ a\sin p,\ bp].$$

2 If the muzzle velocity of a bullet is 600 m s^{-1} and it is ejected horizontally at a height of 2 metres, write down the vector equations of the trajectory and velocity at any time t. Neglect air resistance and leave g in your answers.

(i) Draw the path of the bullet.

(ii) For how long will the bullet be in the air assuming that the ground is level?

(iii) Where will the bullet hit the ground?

(iv) What is the velocity of the bullet at this time?

3 A particle is projected along a smooth plane inclined at an angle θ to the horizontal. If the initial velocity vectors of the particle are $a\mathbf{i}$ and $b\mathbf{j}$ where the axes, both in the plane, are along the horizontal and in the line of greatest slope, respectively, find

(i) the velocity vector at any time t

(ii) the position vector of the particle at its highest point.

If the particle hits a perpendicular stop at its highest point, find the time taken from this stop for the particle to reach again the horizontal line through the point of projection.

4 Find vectors, $\mathbf{r}_1$, $\mathbf{r}_2$, $\mathbf{r}_3$, of the points of intersection of the helix, $\mathbf{h} = 2\cos\theta\,\mathbf{j}+2\sin\theta\,\mathbf{k}+\frac{6\theta}{\pi}\mathbf{i}$, and the straight line, $\mathbf{l} = -4\mathbf{i}+7\mathbf{j}+\sqrt{3}\mathbf{k}+\lambda(-\mathbf{i}+\mathbf{j}+2\mathbf{k})$, where θ and λ are parameters. (Draw a sketch for an approximate solution first.)
If $\mathbf{r}_1$, $\mathbf{r}_2$ and $\mathbf{r}_3$ correspond to vectors of increasing magnitude, show that

$$|2\mathbf{r}_2-(\mathbf{r}_1+\mathbf{r}_3)| = 2(2-\sqrt{3}).$$

5 A point A of a lamina moves with constant acceleration f along a line Ox fixed in the plane of the lamina. The lamina also rotates about an axis through A, perpendicular to its plane, with a constant clockwise angular acceleration, b. At time $t = 0$, the lamina is at rest. Find the relative acceleration of two points on the lamina which are 4 metres apart, on the straight line through A where one of the points is 2 metres directly ahead of A when $t = 1$.

XIII

1 Find vectors of the tangent and normal to the circular helix

$$\mathbf{r} = [a\cos p,\ a\sin p,\ bp]$$

2 Find the radial and transverse components of the velocity and acceleration of a particle moving along the curve $r = a\sin\theta$.

3 A particle P moves in a plane with constant angular speed ω about O. Find $\ddot{r}$ if the rate of increase of its acceleration is parallel to PO.
Find $\ddot{r}$ if the rate of increase of its acceleration is perpendicular to PO.

4 Find the tangential and normal components of the acceleration of the particle moving along the curve $\mathbf{r} = \cos t\,\mathbf{i}+\sin t\,\mathbf{j}+t^2\mathbf{k}$.

5 Find the direction of the normal to the helix $\mathbf{r} = a\cos t\,\mathbf{i}+a\sin t\,\mathbf{j}+bt\mathbf{k}$.

XIV

1 (*a*) A vector of magnitude OP in the direction from O to P is represented by $\mathbf{OP}$. If $\mathbf{OP}-3\mathbf{OQ}+2\mathbf{OR} = \mathbf{0}$, show that P, Q, R are collinear.
(*b*) A unit vector parallel to the x axis is represented by $\mathbf{i}$ and a unit vector parallel to the y axis by $\mathbf{j}$. If $\mathbf{OP} = a\mathbf{i}+s\mathbf{j}$ and $\mathbf{OQ} = -a\mathbf{i}+t\mathbf{j}$, where a is constant and s and t are variables, show that the loci of P and Q are parallel straight lines. In this case find $\mathbf{OQ}$ when $\mathbf{OP} = 2\mathbf{i}+3\mathbf{j}$ and $\mathbf{OQ}$ is perpendicular to $\mathbf{OP}$. (*C*)

2 (i) Points P, Q, R on the sides BC, CA, AB of a triangle ABC are such that

$$\frac{BP}{PC} = \frac{CQ}{QA} = \frac{AR}{RB} = \frac{1+k}{1-k}$$

Prove that the forces $\mathbf{AP}$, $\mathbf{BQ}$, $\mathbf{CR}$ acting in AP, BQ, CR reduce to a couple of moment $2k\Delta$ where Δ is the area of triangle ABC.
(ii) The side AB of a triangle ABC is produced beyond B to P so that $BP = AB$, and the side AC is produced beyond C to Q, so that $CQ = \frac{1}{2}AC$. If a force $k\mathbf{PQ}$ acting in PQ is resolved into components $p\mathbf{BC}$, $q\mathbf{CA}$, $r\mathbf{AB}$ acting in BC, CA, AB, respectively, find p, q, r. (*C*)

3 If $\mathbf{OA} = 2\mathbf{i}+3\mathbf{j}$, $\mathbf{OB} = 3\mathbf{i}-2\mathbf{j}$ and $3\mathbf{OA}+2\mathbf{OB}+\mathbf{OC} = \mathbf{0}$, find the magnitude of $\mathbf{OC}$ and the angle which its direction makes with the positive x axis. (*C*)

4 ABC is a triangle and X, Y, Z are points on the sides BC, CA, AB such that

$$\frac{BX}{XC} = \frac{CY}{YA} = \frac{AZ}{ZB} = k$$

Three forces are represented completely by $\mathbf{AX}$, $\mathbf{BY}$, $\mathbf{CZ}$. Show that the forces are together equivalent to a couple of moment $2\Delta(1-k)/(1+k)$ where Δ is the area of the triangle ABC. (*C*)

5 $ABCD$ is a plane quadrilateral, and forces proportional to the sides act in $\mathbf{AB}$, $\mathbf{AD}$, $\mathbf{CB}$, $\mathbf{CD}$ respectively. Prove that the line of action of the resultant force bisects the line joining the points of intersection (AD, BC) and (AB, CD). Deduce that the middle points of the three diagonals of a complete quadrilateral are collinear. (*C*)

XV

1 P is any point in the plane of a triangle ABC. Show that the resultant of vectors represented by **PA**, **PB** and **PC** is a vector 3**PG**, where G is the centroid of the triangle ABC.
Three points A, B, C on a smooth horizontal table form an equilateral triangle of side a. When a particle of mass m is at any point P on the table it is simultaneously attracted to A by a force $\frac{mg}{a}\mathbf{PA}$, to B by a force $\frac{mg}{a}\mathbf{PB}$ and to C by a force $\frac{mg}{a}\mathbf{PC}$.
Initially it is placed at A and released. Find how long it takes to reach the side BC and its speed as it crosses that side. (*L*)

2 Two forces $\mathbf{F}_1 = \mathbf{i}+\mathbf{j}+\mathbf{k}$ and $\mathbf{F}_2 = \mathbf{i}+2\mathbf{j}-\mathbf{k}$ act through points whose position vectors are $\mathbf{S}_1 = \mathbf{i}+\mathbf{j}+2\mathbf{k}$ and $\mathbf{S}_2 = p\mathbf{j}+5\mathbf{k}$ respectively, relative to a fixed point and three mutually perpendicular unit vectors **i**, **j** and **k**. If the lines of action of $\mathbf{F}_1$ and $\mathbf{F}_2$ intersect, find p, and find the vector equation of the line of action of the resultant of $\mathbf{F}_1$ and $\mathbf{F}_2$. (*L*)

3 With distances measured in nautical miles and speeds in knots, three ships are observed from a coastguard station at half-hour intervals. They have the following distance (**s**) and velocity (**v**) vectors

$\mathbf{s}_1 = 2\mathbf{i}+6\mathbf{j}$ and $\mathbf{v}_1 = 5\mathbf{i}+4\mathbf{j}$ at 12 noon
$\mathbf{s}_2 = 6\mathbf{i}+9\mathbf{j}$ and $\mathbf{v}_2 = 4\mathbf{i}+3\mathbf{j}$ at 12.30 p.m.
$\mathbf{s}_3 = 11\mathbf{i}+6\mathbf{j}$ and $\mathbf{v}_3 = 2\mathbf{i}+7\mathbf{j}$ at 1 p.m.

Prove that if the ships continue with the same velocities, two of them will collide, and find the time of the collision. If at that instant the third ship changes course and then proceeds directly to the scene of collision at its original speed, find at what time it will arrive. (*L*)

4 The two helices $\mathbf{r}_1 = a\cos\theta\,\mathbf{i}+a\sin\theta\,\mathbf{j}+a\theta\mathbf{k}$ and $\mathbf{r}_2 = a\cos n\theta\,\mathbf{i} - a\sin n\theta\,\mathbf{j}+a\theta\mathbf{k}$, where $n > 0$, intersect at successive points P_1, P_2 and P_3. Prove that

$$P_1P_2 = 2a(\sin^2\phi+\phi^2)^{\frac{1}{2}} \quad \text{where} \quad \phi = \pi/(n+1),$$

and find P_1P_3 in terms of a and ϕ. (*L*)

5 A point A of a lamina moves with constant acceleration f along a line Ox fixed in the plane of the lamina, and the lamina also rotates about an axis through A, perpendicular to its plane, with a constant anticlockwise angular acceleration c. At time $t = 0$ the lamina is at rest. Prove that at any later instant of time one point P of the lamina has zero acceleration,

and that if $AP = r$ and the angle $xAP = \theta$, $cr = f\sin\theta$. If $f = \sqrt{3}$, and $r = 2$ when $t = 1$, find the value of c. (*L*)

6 Two identical smooth spheres A and B moving on a horizontal table with velocity vectors $2V\mathbf{i}$ and $V\mathbf{j}$ respectively, collide and the equation of the line of centres at that instant is $\mathbf{r} = \lambda\sin\alpha\mathbf{i} - \lambda\cos\alpha\mathbf{j}$, where λ is a parameter. If the velocity vector of the sphere B after the collision is $u\mathbf{i}$, and the coefficient of restitution between the spheres is $\frac{1}{3}$, prove that $\tan\alpha$ equals either 1 or $\frac{1}{3}$, and in each case find u and the velocity vector of the sphere A after the collision. (*L*)

XVI

1 Three points A, B and C have position vectors $\mathbf{i}+\mathbf{j}+\mathbf{k}$, $\mathbf{i}+2\mathbf{k}$ and $3\mathbf{i}+2\mathbf{j}+3\mathbf{k}$ respectively, relative to a fixed origin O. A particle P starts from B at time $t = 0$, and moves along BC towards C with constant speed 1 unit second^{-1}. Find the position vector of P after t seconds (*a*) relative to O and (*b*) relative to A.

If the angle $PAB = \theta$, find an expression for $\cos\theta$ in terms of t. (*L*)

2 Distances being measured in nautical miles and speeds in knots, a motor boat sets out at 11 a.m. from a position $-6\mathbf{i}-2\mathbf{j}$ relative to a marker buoy, and travels at a steady speed of magnitude $\sqrt{53}$ on a direct course to intercept a ship. The ship maintains a steady velocity vector $3\mathbf{i}+4\mathbf{j}$ and at 12 noon is at a position $3\mathbf{i}-\mathbf{j}$ from the buoy. Find the velocity vector of the motor boat, the time of interception and the position vector of the point of interception from the buoy.

3 If α is a constant such that $0 < \alpha < \frac{1}{2}\pi$, and ϕ is a parameter, prove that the curves

$$\mathbf{r} = b\cos\phi\,\mathbf{i} + b\sin\phi\,\mathbf{j} + b\cos\phi\tan\alpha\,\mathbf{k}$$
$$\mathbf{r} = b\cos\phi\,\mathbf{i} + b\sin\phi\,\mathbf{j} + (2h - b\cos\phi\tan\alpha)\mathbf{k}$$

are both ellipses of eccentricity $\sin\alpha$.

Show that, provided $h^2 < b^2\tan^2\alpha$, the ellipses intersect in two points distant $2(b^2 - h^2\cot^2\alpha)^{\frac{1}{2}}$ apart.

4 Two smooth spheres A and B have masses $2m$ and m respectively, and velocity vectors $3u\mathbf{i}+4u\mathbf{j}$ and $-4u\mathbf{i}+3u\mathbf{j}$ respectively, when they collide with their line of centres parallel to the unit

vector **i**. If the impact causes a loss of energy equal to the original kinetic energy of the sphere B, prove that the coefficient of restitution between the spheres is $\sqrt{(23/98)}$. (*L*)

5 Two particles are projected simultaneously in the same vertical plane, **i** and **j** being unit horizontal and vertical vectors in that plane. The first particle is projected from the origin with velocity vector $nV\cos\alpha\,\mathbf{i}+nV\sin\alpha\,\mathbf{j}$, and the second particle is projected from a position $h\mathbf{i}+k\mathbf{j}$ (where $h > 0$, $k > 0$) with velocity vector $-V\cos\beta\,\mathbf{i}+V\sin\beta\,\mathbf{j}$. Write down the position vectors of each of the particles after time t has elapsed.
Show that the particles cannot collide unless $\sin\beta < n\sin\alpha$, and if they do collide, prove that $\sin(\beta+\gamma) = n(\alpha-\gamma)$, where $\tan\gamma = k/h$. Find the condition imposed on V if the point of collision is above the level of the origin. (*L*)

Answers

Exercise 1

1 (i) $OA = 5$ units $\theta = 53^\circ 8'$
$OB = 13$ units $\theta = 67^\circ 23'$
$AB = 2\sqrt{17}$ units $\theta = 75^\circ 58'$
(ii) $OC = 6\sqrt{2}$ units $\theta = 45^\circ$
$OD = 17$ units $\theta = 61^\circ 56'$
$CD = \sqrt{85}$ units $\theta = 77^\circ 28'$
(iii) $AC = \sqrt{13}$ units $\theta = 66^\circ 42'$
$BD = 3\sqrt{2}$ units $\theta = 45^\circ$
$AD = \sqrt{146}$ units $\theta = 65^\circ 33'$

2 $A_1(9, 12)$ $B_1(1, 2{\cdot}4)$ $OA_1 = 15$ units $\theta = 53^\circ 8'$
$OB_1 = 2{\cdot}6$ units $\theta = 67^\circ 23'$

3 $C_1(3, 3)$ $D_1(16, 30)$ $OC_1 = 3\sqrt{2}$ units $\theta = 45^\circ$
$OD_1 = 34$ units $\theta = 61^\circ 56'$

4 (i) $\frac{1}{2}OA = (1\frac{1}{2}, 2)$ $\frac{1}{2}OB = (2\frac{1}{2}, 6)$ $\frac{1}{2}AB = (4, 8)$
(ii) $\frac{1}{2}OC = (3, 3)$ $\frac{1}{2}OD = (4, 7\frac{1}{2})$ $\frac{1}{2}CD = (7, 10\frac{1}{2})$

5 (i) (8, 16) (ii) (14, 21) (iii) (14, 22) (iv) (22, 37)

6 (i) (6, 8) $\overrightarrow{OE} = 2\overrightarrow{OA}$ (ii) $(2\frac{1}{2}, 6)$ $\overrightarrow{OF} = \frac{1}{2}\overrightarrow{OB}$

7 *OARB* is a parallelogram whose vertex *R* has coordinates (8, 16).

8 *OCSD* is a parallelogram whose vertex *S* has coordinates (14, 21).

9 $(8\frac{1}{2}, 20)$ This number pair represents $8\frac{1}{2}$ kg of apples and 20 oranges.

10 (19, 40, 33) (i) *B* (ii) *D* (iii) *A*

Exercise 2

1 $a = 5$ units $b = 13$ units $c = \sqrt{74}$ units

2 **a** is at 0° **b** is $-22^\circ 37'$ **c** is $-125^\circ 32'$

3

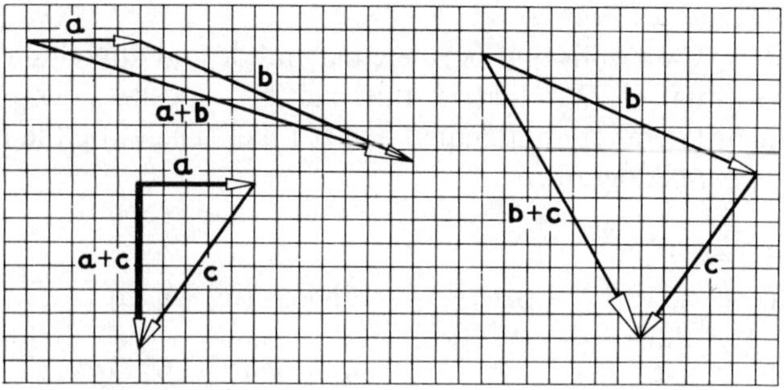

4 (i) $\sqrt{314}$ units (ii) 7 units (iii) $\sqrt{193}$ units

5 (i) $-16°\,23'$ (ii) $-90°$ (iii) $-59°\,45'$

6 (i) (2, 6) (ii) (2, 6) (iii) (4, 3)
(iv) $(-1, -4)$ (v) $(-1, -4)$ (vi) (1, 1)
(vii) $(-3, -1)$ (viii) (0, 1) (ix) (4, 3)

7 (i) 5 units (ii) $\sqrt{5}$ (iii) $\sqrt{13}$ (iv) $\sqrt{2}$

8 (i) $2\sqrt{10}$ (ii) $2\sqrt{10}$ (iii) 5 (iv) $\sqrt{17}$
(v) $\sqrt{17}$ (vi) $\sqrt{2}$ (vii) $\sqrt{10}$ (viii) 1
(ix) 5

9 (i) $53°\,8'$ (ii) $116°\,34'$ (iii) $-123°\,41'$ (iv) $-45°$

10 (i) $71°\,34'$ (ii) $71°\,34'$ (iii) $36°\,52'$ (iv) $-104°\,2'$
(v) $-104°\,2'$ (vi) $45°$ (vii) $-161°\,34'$ (viii) $90°$
(ix) $36°\,52'$

11 $\mathbf{p}+\mathbf{q} = \mathbf{q}+\mathbf{p}$ $\mathbf{p}+\mathbf{s} = \mathbf{s}+\mathbf{p}$ $\mathbf{r}+\mathbf{s} = \mathbf{s}+\mathbf{r}$

Exercise 3

1 (i) (0, 3) (ii) (0, 3) (iii) (3, 5) (iv) (1, 2)
(v) (1, 2)

2 (i) 3 units $90°$ (ii) 3 $90°$ (iii) $\sqrt{34}$ $59°\,2'$
(iv) $\sqrt{5}$ $63°\,26'$ (v) $\sqrt{5}$ $63°\,26'$

3 $\mathbf{c}+\mathbf{a}+\mathbf{b} = \mathbf{b}+\mathbf{c}+\mathbf{a}$ $\mathbf{a}+\mathbf{b}+\mathbf{c}+\mathbf{d} = \mathbf{d}+\mathbf{a}+\mathbf{c}+\mathbf{b}$

4 5 m s^{-1} $36°\,52'$ West of South

5 $2\sqrt{2}$ m s^{-2} NE

6 4·635 newtons $62°\,46'$ East of South

7 $80\sqrt{101}$ km h^{-1} $84°\,17'$ East of South

8 10 km h^{-1} $53°\,8'$ with the down bank

9 1223·2 km h^{-1} $068°\,37'$

10 $3°\,41'$ East of South

11 $1\frac{2}{3}$ m s^{-1} 36° 52′ with the down bank 100 seconds
$133\frac{1}{3}$ metres

12 **AB** = **CD** means that both the magnitude and direction of AB are equal to DC, that is, this pair of opposite sides of the quadrilateral are both equal and parallel. Therefore the figure is a parallelogram.

Exercise 4

1 (i) $[-5, 13]$ (ii) $[-2, -15]$ (iii) $[4, 22]$
(iv) $[4, 22]$ (v) $[0, 8]$

2 (i) $\sqrt{194}$ units 111° 2′ (ii) $\sqrt{229}$ −97° 36′
(iii) $10\sqrt{5}$ 79° 42′ (iv) $10\sqrt{5}$ 79° 42′
(v) 8 90°

3 (i) $[-8, 20]$ (ii) $[-16, 47]$ (iii) $[-12, 39]$
(iv) $[5, -11]$ (v) $[4, 0]$ (vi) $[0, -\frac{2}{3}]$
(vii) $[0, 9]$ (viii) $[-36, 12]$

4 (i) $3[-10, -2] = [-30, -6] = 2[-15, -3]$
(ii) $3[1, 14] = [12, 24]+[-9, 18] = [3, 42]$
(iii) $4[-5, 13] = [-12, 24]-[8, -28] = [-20, 52]$
(iv) $-2[-1, 7] = -[8, 16]-[-10, -2] = [2, -14]$
(v) $7[2, -7] = [10, -35]+[4, -14] = [14, -49]$
(vi) $3[2, -7] = [10, -35]-[4, -14] = [6, -21]$

5 (i) $\sqrt{197} < \sqrt{80}+\sqrt{45}$ that is $14{\cdot}04 < 8{\cdot}944+6{\cdot}708$ (15·652)
(ii) $\sqrt{53} > \sqrt{80}-\sqrt{45}$ that is $7{\cdot}28 > 8{\cdot}944-6{\cdot}708$ (2·236)

6 5 N parallel to the x axis and $5\sqrt{3}$ N parallel to the y axis.

7 1·88 N, 243° 22′

8 Without the λ, μ theorem:

AB+**BM** = **AM**

AC+**CM** = **AM**

Add **AB**+**AC**+**BM**+**CM** = 2**AM**

Since M is the mid-point of BC,

BM+**CM** = **0**

Hence **AB**+**AC** = 2**AM**

With the λ, μ theorem:

1**AB**+1**AC** = (1+1)**AM**

where $\lambda = \mu = 1$ and M is the mid-point of BC.

Hence **AB**+**AC** = 2**AM**

9 Using the figure shown here, we have

DA+**AE** = **DE** or 2**DA**+2**AE** = 2**DE**

Again **BA**+**AC** = **BC**

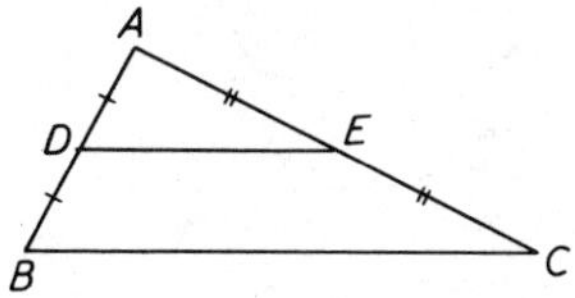

But $\quad 2\mathbf{DA} = \mathbf{BA} \quad$ and $\quad 2\mathbf{AE} = \mathbf{AC}$

Hence $\quad 2\mathbf{DE} = \mathbf{BC}$

that is *DE* is equal to one-half *BC* and is parallel to *BC*.

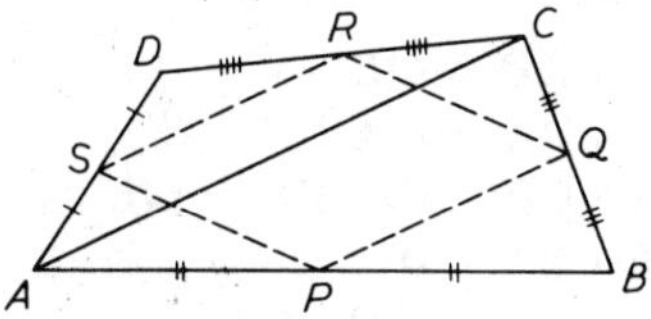

10 From the solution to question 9

in $\Delta\ ADC$, $\quad \mathbf{SR} = \frac{1}{2}\mathbf{AC}$

in $\Delta\ ABC$, $\quad \mathbf{PQ} = \frac{1}{2}\mathbf{AC}$

Hence $\quad \mathbf{SR} = \mathbf{PQ}$

that is the figure *PQRS* is a parallelogram.

11

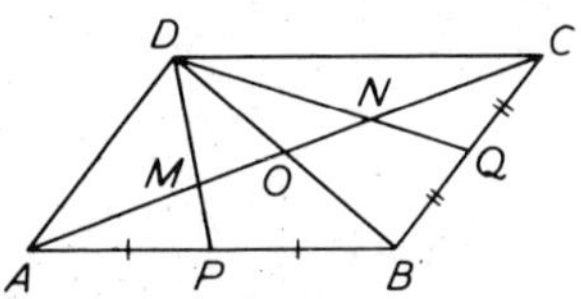

In the parallelogram *ABCD* let *P* and *Q* be the mid-points of *AB* and *BC* respectively, and *M* and *N* be the points of trisection of *DP* and *DQ*. From the λ, μ theorem, in $\Delta\ ADP$

$$\mathbf{AM} = \tfrac{1}{3}\mathbf{AD} + \tfrac{2}{3}\mathbf{AP}$$

whence $\quad 3\mathbf{AM} = \mathbf{AD} + 2\mathbf{AP} = \mathbf{AD} + \mathbf{AB}$

But $\quad \mathbf{AC} = \mathbf{AD} + \mathbf{AB}$

Hence $\quad 3\mathbf{AM} = \mathbf{AC} \quad$ and $\quad \mathbf{AM} = \frac{1}{3}\mathbf{AC}$

Therefore *M* trisects *AC*. Similarly, use *CN* in $\Delta\ CDQ$ to show *N* trisects *CA*.

12 Using the figure shown in the answer to question 10, we can either use solutions based on that question and on question 9, or the following:

$$2\mathbf{QC} + 2\mathbf{CR} = \mathbf{BD} = 2\mathbf{PA} + 2\mathbf{AS}$$

Therefore $\quad \mathbf{QC} + \mathbf{CR} = \mathbf{PA} + \mathbf{AS}$,

whence $\quad \mathbf{QR} = \mathbf{PS}$

that is *PQRS* is a parallelogram and so the diagonals, *PR* and *SQ*, bisect each other.

13

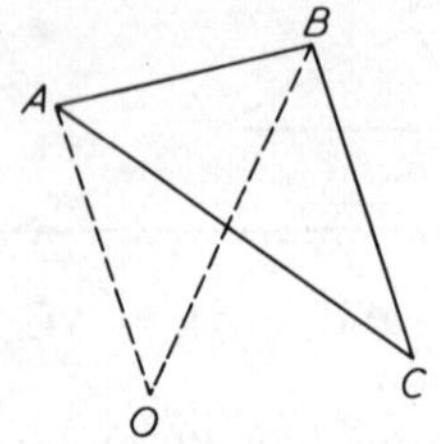

From the diagram,

$$\mathbf{OA}+\mathbf{AB} = \mathbf{OB}$$
$$\mathbf{CB}+\mathbf{BA} = \mathbf{CA}$$

Hence $\quad \mathbf{OA}+\mathbf{CB}+\mathbf{AB}+\mathbf{BA} = \mathbf{OB}+\mathbf{CA}$

But $\quad \mathbf{AB}+\mathbf{BA} = \mathbf{0}$

Therefore $\quad \mathbf{OA}+\mathbf{CB} = \mathbf{OB}+\mathbf{CA}$

14

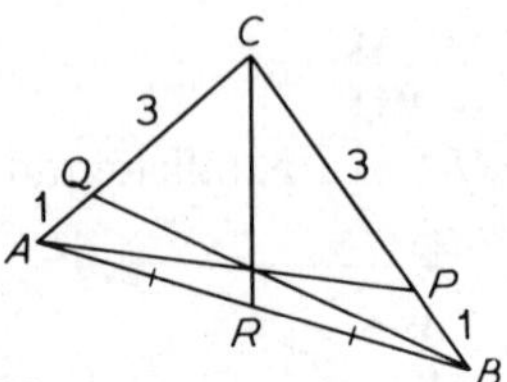

From the diagram

$$\lambda\mathbf{AB}+\mu\mathbf{AC} = (\lambda+\mu)\mathbf{AP}$$

where $\mu:\lambda = 1:3$, that is $3\mu = \lambda$.

Hence $\quad 3\mathbf{AB}+\mathbf{AC} = 4\mathbf{AP}$

Similarly $\quad \mathbf{BC}+3\mathbf{BA} = 4\mathbf{BQ}$

and $\quad \mathbf{CA}+\mathbf{CB} = 2\mathbf{CR}$

Hence $\quad 3\mathbf{AB}+\mathbf{AC}+\mathbf{BC}+3\mathbf{BA}+\mathbf{CA}+\mathbf{CB} = 4\mathbf{AP}+4\mathbf{BQ}+2\mathbf{CR}$

whence $\quad 2\mathbf{AP}+2\mathbf{BQ}+\mathbf{CR} = \mathbf{0}$

Exercise 5

1 $[4, -1]$

2 $[1, -2]$

3 $\mathbf{OM} = [2, 2] \qquad \mathbf{AM} = [2, -2]$

4 Since $AP:PB = m:n$, then $n(\mathbf{p}-\mathbf{a}) = m(\mathbf{b}-\mathbf{p})$, that is $\mathbf{p}(m+n) = m\mathbf{b}+n\mathbf{a}$, whence the result.

5 Either, since the sum of the coefficients, $3+4-7$, is zero, therefore, the points A, B and C are collinear or $3\mathbf{OA}+4\mathbf{OB} = 7\mathbf{ON}$ where $AN:NB = 4:3$ and the points A, N and B are collinear; but this is identical to the given equation so that N coincides with C and the points A, B and C are collinear.

6 If O is the mid-point of BD, then

$$\mathbf{AB}+\mathbf{AD} = 2\mathbf{AO}$$

Similarly $\mathbf{CB}+\mathbf{CD} = 2\mathbf{CO}$

Therefore $\mathbf{AB}+\mathbf{AD}+\mathbf{CB}+\mathbf{CD} = 2(\mathbf{AO}+\mathbf{CO})$

The condition is $\mathbf{AO}+\mathbf{CO} = \mathbf{0}$.

7 Since $\mathbf{AB}+\mathbf{BC} = \mathbf{AC}$ and $\mathbf{AE}+\mathbf{ED} = \mathbf{AD}$, then

$$\mathbf{AB}+\mathbf{BC}+\mathbf{AC}+\mathbf{AE}+\mathbf{ED}+\mathbf{AD} = 2\mathbf{AC}+2\mathbf{AD} = 4\mathbf{AM}$$

where M is the mid-point of CD.

8 From $\Delta\ ABC$, $\mathbf{AB}+\mathbf{AC} = 2\mathbf{AN}$ or

$$\mathbf{AB}+\mathbf{AC}+\mathbf{BC} = (\mathbf{AB}+\mathbf{BC})+\mathbf{AC} = 2\mathbf{AC} = 4\mathbf{MN}$$

from $\Delta\ ANB$, $2\mathbf{AN}+2\mathbf{BN} = 4\mathbf{MN}$

whence $\mathbf{AB}+\mathbf{AC}+\mathbf{BC} = 4\mathbf{MN}$

9 As in the previous solutions, $\mathbf{AB}+\mathbf{AC} = 2\mathbf{AN}$ and $2\mathbf{AN}+2k\mathbf{BN} = 2(k+1)\mathbf{MN}$ where $\mathbf{AM}:\mathbf{MB} = k:1$.

Therefore $\mathbf{AB}+\mathbf{AC}+k\mathbf{BC} = 2(k+1)\mathbf{MN}$.

10 173·94 km at 043° 18′

11 4·617 nautical miles at 081° 4′

12 9·174 knots at 202° 40′

13 1044 km h^{-1} at 215° 30′

14 $2\sqrt{13}$ km h^{-1} from 213° 41′

15
$$_S g_O = \frac{m_1\mathbf{a}_1+m_2\mathbf{a}_2+\ldots+m_n\mathbf{a}_n}{m_1+m_2+\ldots+m_n}$$

Relative to another fixed point P, vector $\mathbf{p}$, the centroid is $\mathbf{g}'$, say, where $_S g'_P = {_S g'_O} - {_P g'_O} = \mathbf{g}-\mathbf{p}$

By definition, however,

$$_S g'_P = \frac{m_1(\mathbf{a}_1-\mathbf{p})+m_2(\mathbf{a}_2-\mathbf{p})+\ldots+m_n(\mathbf{a}_n-\mathbf{p})}{m_1+m_2+\ldots+m_n} = \mathbf{g}-\mathbf{p}$$

whence the result.

16 Let
$$\mathbf{g} = \frac{m_1\mathbf{a}_1+m_2\mathbf{a}_2+\ldots+m_n\mathbf{a}_n}{m_1+m_2+\ldots+m_n}$$

and
$$\mathbf{g}' = \frac{m'_1\mathbf{a}'_1+m'_2\mathbf{a}'_2+\ldots+m'_n\mathbf{a}'_n}{m'_1+m'_2+\ldots+m'_n}$$

The centroid of the composite system is $\mathbf{g}''$, say, where

$$\mathbf{g}'' = \frac{m_1\mathbf{a}_1 + m_2\mathbf{a}_2 + \ldots + m_n\mathbf{a}_n + m_1'\mathbf{a}_1' + m_2'\mathbf{a}_2' + \ldots + m_n'\mathbf{a}_n'}{m_1 + m_2 + \ldots + m_n + m_1' + m_2' + \ldots + m_n'}$$

But $\mathbf{g}'' = \dfrac{Mg + M'g'}{M + M'}$,

where M and M' are the total masses of S and S',

$$= \frac{m_1\mathbf{a}_1 + m_2\mathbf{a}_2 + \ldots + m_n\mathbf{a}_n + m_1'\mathbf{a}_1' + m_2'\mathbf{a}_2' + \ldots + m_n'\mathbf{a}_n'}{m_1 + m_2 + \ldots + m_n + m_1' + m_2' + \ldots + m_n'}$$

whence the result.

Exercise 6

1 (i) $5\mathbf{i}+2\mathbf{j}$ (ii) $3\mathbf{i}+4\mathbf{j}$ (iii) $9\mathbf{i}+5\mathbf{j}$ (iv) $7\mathbf{j}$
(v) $7(\mathbf{i}+2\mathbf{j})$

2 $p = 5$ units and $q = \sqrt{2}$ units

3 $\mathbf{p}$ makes an angle of $36° 52'$ and $\mathbf{q}$, $135°$

4 $|\mathbf{i}+\mathbf{j}| = \sqrt{1^2+1^2} = \sqrt{2}$ units
$|\mathbf{i}-\mathbf{j}| = \sqrt{1^2+(-1)^2} = \sqrt{2}$ units

5 (i) 5 units (ii) 5 (iii) $\sqrt{41}$ (iv) $\sqrt{13}$ (v) $\sqrt{5}$

6 (i) $45°$ (ii) $-45°$ (iii) $-135°$ (iv) $135°$
(v) $63° 26'$ (vi) $116° 34'$ (vii) $56° 19'$
(viii) $-56° 19'$ (ix) $123° 41'$ (x) $-123° 41'$

7 (i) $4\mathbf{i}+\mathbf{j}-\mathbf{k}$ (ii) $3\mathbf{i}-3\mathbf{j}+8\mathbf{k}$ (iii) $-8\mathbf{i}-5\mathbf{j}+\mathbf{k}$
(iv) $\mathbf{i}-\mathbf{j}$ (v) $6\mathbf{i}-\mathbf{j}+7\mathbf{k}$ (vi) $-6\mathbf{i}-6\mathbf{j}+4\mathbf{k}$

8 $\mathbf{a}$: $l = 2/\sqrt{14}$ $m = -1/\sqrt{14}$ $n = 3/\sqrt{14}$
$\mathbf{b}$: $l = 1/\sqrt{6}$ $m = 1/\sqrt{6}$ $n = -2/\sqrt{6}$
$\mathbf{c}$: $l = -3/\sqrt{14}$ $m = -2/\sqrt{14}$ $n = 1/\sqrt{14}$

9 $|\mathbf{i}+\mathbf{j}+\mathbf{k}| = \sqrt{1^2+1^2+1^2} = \sqrt{3}$ units
$|\mathbf{i}-\mathbf{j}-\mathbf{k}| = \sqrt{1^2+(-1)^2+(-1)^2} = \sqrt{3}$
$|\mathbf{i}+\mathbf{j}-\mathbf{k}| = \sqrt{1^2+1^2+(-1)^2} = \sqrt{3}$

10 (i) 3 units (ii) 3 (iii) 6 (iv) 9 (v) 9

11 $l = 1/\sqrt{3}$ $m = 1/\sqrt{3}$ $n = 1/\sqrt{3}$
$l^2+m^2+n^2 = \frac{1}{3}+\frac{1}{3}+\frac{1}{3} = 1$

12 (i) $l = 1/9$ $m = 4/9$ $n = 8/9$
(ii) $l = 1/9$ $m = -4/9$ $n = 8/9$
(iii) $l = 4/9$ $m = -4/9$ $n = -7/9$
(iv) $l = 1/\sqrt{14}$ $m = 2/\sqrt{14}$ $n = 3/\sqrt{14}$
(v) $l = -2/3$ $m = 1/3$ $n = -2/3$

13 (i) $\frac{1}{9}(\mathbf{i}+4\mathbf{j}+8\mathbf{k})$ (ii) $\frac{1}{9}(\mathbf{i}-4\mathbf{j}+8\mathbf{k})$

(iii) $\frac{1}{9}(4\mathbf{i}-4\mathbf{j}-7\mathbf{k})$ (iv) $\frac{1}{\sqrt{14}}(\mathbf{i}+2\mathbf{j}+3\mathbf{k})$

(v) $\frac{1}{3}(-2\mathbf{i}+\mathbf{j}-2\mathbf{k})$

14 $3\mathbf{i}+4\mathbf{j}$ 5 units $53°\,8'$

15 $-3\mathbf{i}-4\mathbf{j}$ 5 units $-126°\,52'$

16 $-5\mathbf{i}+12\mathbf{j}$ 13 units $112°\,37'$

17 $5\mathbf{i}-12\mathbf{j}$ 13 units $-67°\,23'$

18 $-2\mathbf{i}+\mathbf{j}-8\mathbf{k}$ $\sqrt{69}$ units

$l = -2/\sqrt{69}$ $m = 1/\sqrt{69}$ $n = -8/\sqrt{69}$

19 $-2\mathbf{i}-3\mathbf{j}+2\mathbf{k}$ $\sqrt{17}$ units

$l = -2/\sqrt{17}$ $m = -3/\sqrt{17}$ $n = 2/\sqrt{17}$

20 (*a*) $2\mathbf{i}-\mathbf{j}+8\mathbf{k}$ $\sqrt{69}$ units

$l = 2/\sqrt{69}$ $m = -1/\sqrt{69}$ $n = 8/\sqrt{69}$

(*b*) $2\mathbf{i}+3\mathbf{j}-2\mathbf{k}$ $\sqrt{17}$ units

$l = 2/\sqrt{17}$ $m = 3/\sqrt{17}$ $n = -2/\sqrt{17}$

21 $|\hat{\mathbf{a}}| = 1$ unit at $53°\,8'$ $|\hat{\mathbf{b}}| = 1$ unit

$l = \frac{1}{3}$ $m = -\frac{2}{3}$ $n = \frac{2}{3}$

22 From the diagram we see that the mid-point of AB is

$$\mathbf{p} = \tfrac{1}{2}(\mathbf{a}+\mathbf{b})$$

and the mid-point of CD $\mathbf{r} = \frac{1}{2}(\mathbf{c}+\mathbf{d})$

If G is the mid-point of PR, $\mathbf{g} = \frac{1}{4}(\mathbf{a}+\mathbf{b}+\mathbf{c}+\mathbf{d})$.

Since this expression is symmetrical for the position vectors of A, B, C and D, we can, therefore, say that G is also the mid-

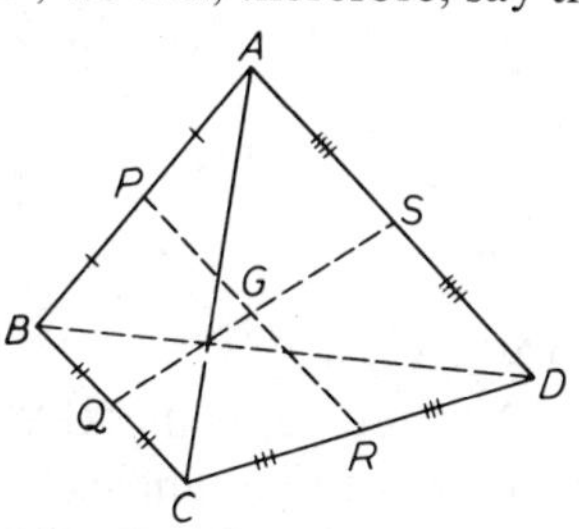

point of both the other line segments joining the mid-points of opposite edges.

23 Either use the properties of question 22 and its solution or the following: let the centroid of the triangle be at H, then

$$\mathbf{h} = \frac{\mathbf{b}+\mathbf{c}+\mathbf{d}}{3}$$

The centroid of the tetrahedron now lies on AH. Consider the point G such that $AG{:}GH = 3{:}1$, then

$$\mathbf{g} = \frac{\mathbf{a}+3\dfrac{\mathbf{b}+\mathbf{c}+\mathbf{d}}{3}}{4} = \tfrac{1}{4}(\mathbf{a}+\mathbf{b}+\mathbf{c}+\mathbf{d}),$$

whence the result.

24 [0, −2, −1]

25 [1, 2, −1]

26 [1, 2, −2]

27 After 1 second: $\mathbf{s}_1 = 40\mathbf{i}$ $\mathbf{s}_2 = 30\mathbf{i}$
After 2 seconds: $\mathbf{s}_1 = 70\mathbf{i}$ $\mathbf{s}_2 = 70\mathbf{i}$
After 3 seconds: $\mathbf{s}_1 = 100\mathbf{i}$ $\mathbf{s}_2 = 110\mathbf{i}$
The second particle overtakes the first at time t where t is given by the equation

$$10\mathbf{i}+(30\mathbf{i})t = -10\mathbf{i}+(40\mathbf{i})t$$

whence $t = 2$ seconds as we can see from the short table compiled already.

28 At noon, the distance is 25 metres; at 1 second later, $5\sqrt{2}$ metres and 2 seconds later, $5\sqrt{5}$ metres.

29 At 2 p.m., $\mathbf{s}_1 = 11\mathbf{i}+21\mathbf{j}$ and $\mathbf{s}_2 = 8\mathbf{i}-11\mathbf{j}$. Distance $\sqrt{1033}$ nautical miles.

30	$\mathbf{i}-\mathbf{j}$	$\sqrt{2}$ m s^{-1}	135°
31	$4\mathbf{i}-12\mathbf{j}$	$4\sqrt{10}$ units	108° 26′
32	$3\mathbf{i}-4\mathbf{j}$	5 units	126° 52′
33	$8\mathbf{i}+6\mathbf{j}$		

34 $\mathbf{v}_A = [1, 2, -2]$ $\mathbf{v}_C = [1, 2, -2]$ $|\mathbf{v}_A| = |\mathbf{v}_C| = 3$
$\mathbf{v}_A = \mathbf{v}_C$

Exercise 7

1 $\mathbf{ON} = [2, 2]$ $\mathbf{AN} = [2, -2]$

(*a*) $\mathbf{ON}.\mathbf{AN} = 2\times 2+2\times(-2) = 0 = ON.AN\cos\theta$. Therefore $\cos\theta = 0$ and so N is a right angle.

(*b*) $AN^2+ON^2 = 8+8 = 16 = OA^2$. Therefore N is a right angle.

2 Either $\mathbf{a}.\mathbf{b} = 0$, whence $a_1b_1+a_2b_2 = 0$ immediately, or when $\mathbf{OA}^2+\mathbf{AB}^2 = \mathbf{OB}^2$, then $a_1^2+a_2^2+(b_1-a_1)^2+(b_2-a_2)^2 = b_1^2+b_2^2$ and, again, $a_1b_1+a_2b_2 = 0$.

3 $a_1b_1+a_2b_2+a_3b_3 = 0$

4

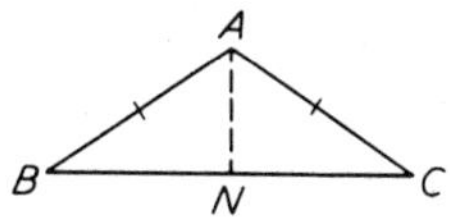

In the diagram, let **AB** = **b** and **AC** = **c**, then

$$\mathbf{AN}.\mathbf{BC} = \tfrac{1}{2}(\mathbf{b}+\mathbf{c}).(\mathbf{c}-\mathbf{b})$$
$$= \tfrac{1}{2}(\mathbf{b}.\mathbf{c}+\mathbf{c}.\mathbf{c}-\mathbf{b}.\mathbf{b}-\mathbf{c}.\mathbf{b})$$
$$= \mathbf{c}^2-\mathbf{b}^2$$

But $|\mathbf{c}| = |\mathbf{b}|$, therefore, $\mathbf{AN}.\mathbf{BC} = 0$, whence the result.

5

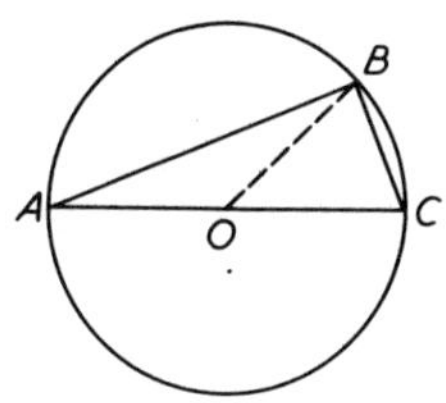

$$\mathbf{AB}.\mathbf{BC} = (\mathbf{b}-\mathbf{a}).(\mathbf{c}-\mathbf{b})$$
$$= \mathbf{b}.\mathbf{c}-\mathbf{a}.\mathbf{c}-\mathbf{b}.\mathbf{b}+\mathbf{a}.\mathbf{b} \qquad \ldots (1)$$

Now $\mathbf{a} = -\mathbf{c}$ and $|\mathbf{a}| = |\mathbf{b}| = |\mathbf{c}|$ ... (2)

Simplifying (1), therefore, we have

$$\mathbf{b}.\mathbf{c}+\mathbf{c}.\mathbf{c}-\mathbf{b}.\mathbf{b}-\mathbf{c}.\mathbf{b} = \mathbf{c}^2-\mathbf{b}^2 = 0 \quad \text{from (2).}$$

Therefore $\mathbf{AB}.\mathbf{BC} = 0$, whence the result.

6 Let the altitudes *AM* and *BN* intersect at *O*. Therefore

$$\mathbf{OA}.\mathbf{BC} = 0 \quad \text{and} \quad \mathbf{OB}.\mathbf{CA} = 0 \qquad \ldots (1)$$

We now have to prove that *COP* is perpendicular to *AB*.

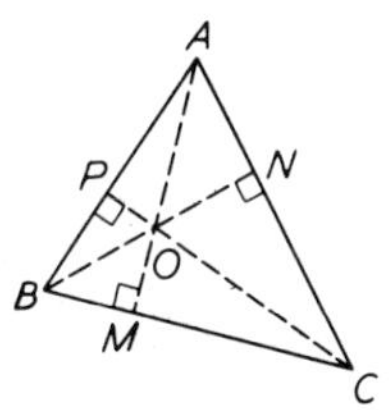

From (1) $\mathbf{a}.(\mathbf{c}-\mathbf{b}) = 0$ and $\mathbf{b}.(\mathbf{a}-\mathbf{c}) = 0$

that is $\mathbf{a}.\mathbf{c} = \mathbf{a}.\mathbf{b}$ and $\mathbf{b}.\mathbf{a} = \mathbf{b}.\mathbf{c}$

Since $\mathbf{a}.\mathbf{b} = \mathbf{b}.\mathbf{a}$, then $\mathbf{a}.\mathbf{c} = \mathbf{b}.\mathbf{c}$

that is $\mathbf{c}.(\mathbf{a}-\mathbf{b}) = 0$ but this is $\mathbf{OC}.\mathbf{AB} = 0$,

which means that *CP* is perpendicular to *AB*, whence the result.

7 20 J
8 15 J
9 8 J
10 26 J
11 26 J
12 20 J
13 4 J

14 (i) -1 (ii) $\hat{\mathbf{a}} = \frac{1}{\sqrt{3}}(\mathbf{i}+\mathbf{j}+\mathbf{k})$ $\hat{\mathbf{b}} = \frac{1}{3}(2\mathbf{i}-\mathbf{j}-2\mathbf{k})$

(iii) **a**: $a_x = 1$ $a_y = 1$ $a_z = 1$
$l = 1/\sqrt{3}$ $m = 1/\sqrt{3}$ $n = 1/\sqrt{3}$
b: $b_x = 2$ $b_y = -1$ $b_z = -2$
$l = 2/3$ $m = -1/3$ $n = -2/3$

15 $\mathbf{p}.(\mathbf{q}+\mathbf{r}) = -6$ $\mathbf{p}.\mathbf{q} = 5$ $\mathbf{p}.\mathbf{r} = -11$ $5+(-11) = -6$
16 (i) $5(2\mathbf{i}-3\mathbf{j}+\mathbf{k})$ (ii) $-\mathbf{i}+2\mathbf{j}-3\mathbf{k}$ (iii) $9\mathbf{i}-13\mathbf{j}+2\mathbf{k}$
17 $\cos\theta = 1/3$, so that $\theta = 70^\circ\, 32'$
18 $A = 45^\circ$, $B = 90^\circ$ and $C = 45^\circ$
19

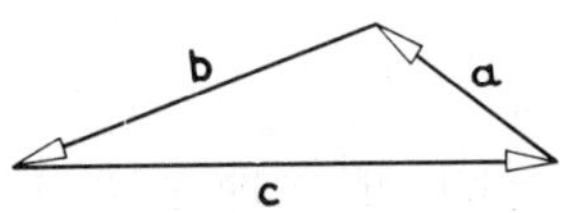

From the diagram, $-\mathbf{a} = \mathbf{b}+\mathbf{c}$
so that $\mathbf{a}^2 = \mathbf{b}^2+\mathbf{c}^2+2\mathbf{b}.\mathbf{c}$ (cosine rule)
20 $\mathbf{a}.\mathbf{b} = \cos\theta\cos\phi+\sin\theta\sin\phi = ab\cos(\theta-\phi)$
Since $|\mathbf{a}| = |\mathbf{b}| = 1$, we can obtain the result.
21 No: therefore, no vector cancellation law exists.

Exercise 8

1 (i) $-6\mathbf{i}+6\mathbf{j}-2\mathbf{k}$ (ii) $\mathbf{i}-\mathbf{j}+\mathbf{k}$ (iii) $4\mathbf{i}-8\mathbf{j}+4\mathbf{k}$
2 (i) $\mathbf{b}\times\mathbf{a} = 6\mathbf{i}-6\mathbf{j}+2\mathbf{k}$ (ii) $\mathbf{c}\times\mathbf{b} = -\mathbf{i}+\mathbf{j}-\mathbf{k}$
(iii) $\mathbf{c}\times\mathbf{a} = -4\mathbf{i}+8\mathbf{j}-4\mathbf{k}$
Therefore the given equalities hold in every case.
3 (i) $-2\mathbf{i}-2\mathbf{j}+2\mathbf{k}$ (ii) $2\mathbf{i}+2\mathbf{j}-2\mathbf{k}$
4 $-18\mathbf{i}+18\mathbf{j}-6\mathbf{k}$
5 $10\mathbf{i}+4\mathbf{j}-6\mathbf{k}$
6 $a = \sqrt{10}$ $b = \sqrt{20}$ $c = \sqrt{10}$ Area = 5 units^2
7 $A = 107^\circ\, 28'$ $B = C = 36^\circ\, 16'$
$a = \sqrt{14}$ $b = c = \sqrt{10}$ Area = 4·77 units^2
8 $\pm\frac{1}{3}[1, -2, 2]$

9 $\frac{1}{2}[1, \pm 1, \sqrt{2}]$

10 $5\sqrt{2}/2$ units^2 $\quad 5\sqrt{2}$ units^2 $\quad$ No

11 $2\sqrt{2}/3$

12 **a** is parallel to **b**, $\mathbf{a} = 0$ or $\mathbf{b} = 0$, $\mathbf{a} = k\mathbf{b}$, **a** and **b** are collinear. Equation of line through P_1 with direction numbers a_x, a_y, a_z where $\mathbf{P_1P} \times \mathbf{a} = \mathbf{0}$

13 $|\mathbf{p} \times \mathbf{q}|^2 = (p_yq_z - q_yp_z)^2 + (p_xq_z - q_xp_z)^2 + (p_xq_y - q_xp_y)^2 = \ldots$

$$|\mathbf{p}|^2|\mathbf{q}|^2 - (\mathbf{p}.\mathbf{q})^2 = (p_x^2 + p_y^2 + p_z^2)(q_x^2 + q_y^2 + q_z^2) - (p_xq_x + p_yq_y + p_zq_z)^2 = \ldots$$

whence the result follows after simplification.

14 (i) $6\mathbf{i} + 8\mathbf{j} + 6\mathbf{k}$ $\quad$ (ii) -80 $\quad$ (iii) -14

15 $-\frac{5}{11}(81\mathbf{i} + 3\mathbf{j} - 16\mathbf{k})$

16 $3\mathbf{i} - 12\mathbf{j} + 9\mathbf{k}$

Speed is $3\sqrt{26}$ and $l = \dfrac{\sqrt{26}}{26}$ $\quad m = \dfrac{-2\sqrt{26}}{13}$ $\quad n = \dfrac{3\sqrt{26}}{26}$

Exercise 9

1 $\mathbf{a} - \mathbf{b} = 2\mathbf{i} + 3\mathbf{j}$ which is a vector parallel to the x, y plane.

2 $\mathbf{p} - \mathbf{q} = 2\mathbf{j} + 2\mathbf{k}$ which is a vector parallel to the y, z plane.

3 (i) For $\mathbf{r} = [x, y, z]$, $(\mathbf{r} - \mathbf{a}).\hat{\mathbf{n}} = 0$
(ii) $x - 2y - 3z = 9$
(iii) $2x + 2y + z = 11$

4 (i) $45°$ $\quad$ (ii) $33° 34'$

5 $3x - 4y - 3z = 6$

6 $|\hat{\mathbf{n}}.\mathbf{QP}|$ gives the projection of QP on to the unit normal, that is $1.QP\cos\theta$, and so gives the perpendicular distance of P from the line.

7 1

8 (i) $4/3$ $\quad$ (ii) $2/\sqrt{11}$

9 (i) $x + z = 1$ $\quad$ (ii) $x - y - 3z = 11$

10 $x^2 + y^2 + z^2 - 6x - 4y - 10z + 29 = 0$

Exercise 10

1 $\mathbf{r} = \mathbf{i} + 2\mathbf{j} + 3\mathbf{k} + t(2\mathbf{i} - 4\mathbf{j} + 7\mathbf{k})$ or $\mathbf{r} = [1 + 2t, 2 - 4t, 3 + 7t]$

or $\dfrac{x-1}{2} = \dfrac{y-2}{-4} = \dfrac{z-3}{7} = t$

2 One form is $\mathbf{r} = [3+4t,\ -2+6t,\ 5-7t]$

3 (i) $\mathbf{r} = [a_1+t(b_1-a_1), a_2+t(b_2-a_2)]$

(ii) $\mathbf{r} = [2+3t, 5-6t]$

(iii) $\mathbf{r} = [-4+7t, t]$

(iv) $\mathbf{r} = [7-13t,\ -2+6t]$

4 (i) $\mathbf{r} = [a_1+t(b_1-a_1), a_2+t(b_2-a_2), a_3+t(b_3-a_3)]$

(ii) $\mathbf{r} = [3-4t, 2, 1]$

(iii) $\mathbf{r} = [1+2t, 4t, -2+t]$

(iv) $\mathbf{r} = [t, 2-3t, 2t]$

5 (i)

$$\frac{x-\begin{vmatrix}b_4 & a_4\\ b_2 & a_2\end{vmatrix}\div\begin{vmatrix}a_1 & b_1\\ a_2 & b_2\end{vmatrix}}{\begin{vmatrix}a_2 & b_2\\ a_3 & b_3\end{vmatrix}} = \frac{y-\begin{vmatrix}a_1 & b_1\\ a_4 & b_4\end{vmatrix}\div\begin{vmatrix}a_2 & b_2\\ a_1 & b_1\end{vmatrix}}{\begin{vmatrix}a_3 & b_3\\ a_1 & b_1\end{vmatrix}} = \frac{z}{\begin{vmatrix}a_1 & b_1\\ a_2 & b_2\end{vmatrix}}$$

(ii) One form is: $\mathbf{r} = [-\frac{1}{4}, 1, 4\frac{1}{2}]+t[0, 4, 12]$

(iii) $\mathbf{r} = [-1, 6, 3]+t[1, -7, -4]$

(iv) $\mathbf{r} = [-1, 1, 1]+t[-1, 1, 1]$

6 5 part (ii) This line does not cut the plane $x = 0$

For $y = 0,\ x = -\frac{1}{4}$ and $z = 1\frac{1}{2}$

For $z = 0,\ x = -\frac{1}{4}$ and $y = -\frac{1}{2}$

5 part (iii) $x = 0,\ y = -1,\ z = -1$

$y = 0,\ x = -\frac{1}{7},\ z = -\frac{3}{7}$

$z = 0,\ x = -\frac{1}{4},\ y = \frac{3}{4}$

5 part (iv) $x = 0,\ y = 0,\ z = 0$

7 5/9

8 $\mathbf{r} = [1, 3, 6]+t[-1, 4, 6]$

9 (i) 0 (ii) $35/\sqrt{195}$

10 $\mathbf{r} = \mathbf{p}+t[-3, 3, 3]$ and one form is $\dfrac{3x-7}{-9} = \dfrac{3y+5}{9} = \dfrac{z}{3}$

Exercise 11

1 $\mathbf{AP}^2+\mathbf{BP}^2 = \mathbf{AB}^2$

2 (i) $\mathbf{r} = 2\cos\theta\,\mathbf{i}+2\sin\theta\,\mathbf{j}$

(ii) $\mathbf{r} = (2+4\cos\theta)\mathbf{i}+(3+4\sin\theta)\mathbf{j}$

3 (i) A circle centre (0, 0) and radius 1 unit.

(ii) A circle (0, 0) and radius 3 units.

4 (i) Centre (1, 2), radius 1 unit

(ii) Centre (−3, 0), radius 2 units

(iii) Centre (1, 0), radius 3 units

5 (i) $x^2+y^2-2x-4y+4 = 0$
(ii) $x^2+y^2+6x+5 = 0$
(iii) $x^2+y^2-2x-8 = 0$

6 (i) $\mathbf{r} = a\mathbf{i}$
(ii) $\mathbf{r} = a\sqrt{2}(\mathbf{i}+\mathbf{j})/2$
(iii) $\mathbf{r} = a\mathbf{j}$
(iv) $\mathbf{r} = a(-\mathbf{i}+\sqrt{3}\mathbf{j})/2$
(v) $\mathbf{r} = a(\mathbf{i}-\sqrt{3}\mathbf{j})/2$

7 $\mathbf{r} = 2\cos\frac{2}{3}\pi t\,\mathbf{i}+2\sin\frac{2}{3}\pi t\,\mathbf{j}$
(i) $\mathbf{r} = \mathbf{i}+\sqrt{3}\mathbf{j}$
(ii) $\mathbf{r} = -\sqrt{3}\mathbf{i}+\mathbf{j}$

8 $t = 0$ or 2 Intercept is $2\sqrt{2}$ units

9 $p = 107°\,6'$ or $27°\,40'$

10 $p = \pm\sqrt{3}/3$

11 The scalar product is $\mathbf{r}_1 . \mathbf{r}_2 = r^2 = 1$ in this case. $\theta = \pm\pi/3$ and $t = \sqrt{10}(3+\sqrt{3})/2$

Exercise 12

1 (i)

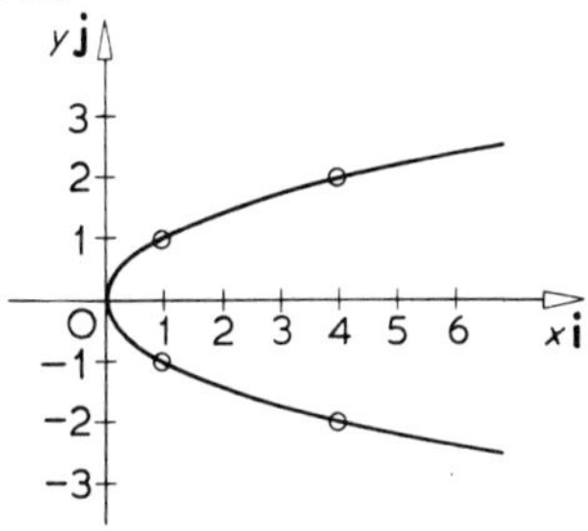

(ii)

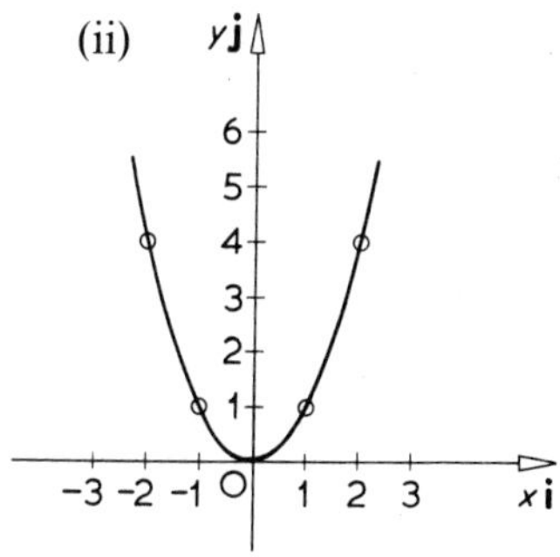

(iii)

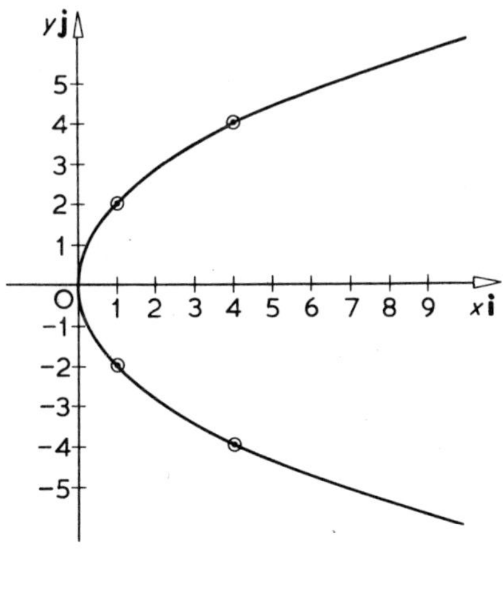

2 (i) $y^2 = x$
(ii) $y = x^2$
(iii) $y^2 = 4x$

3

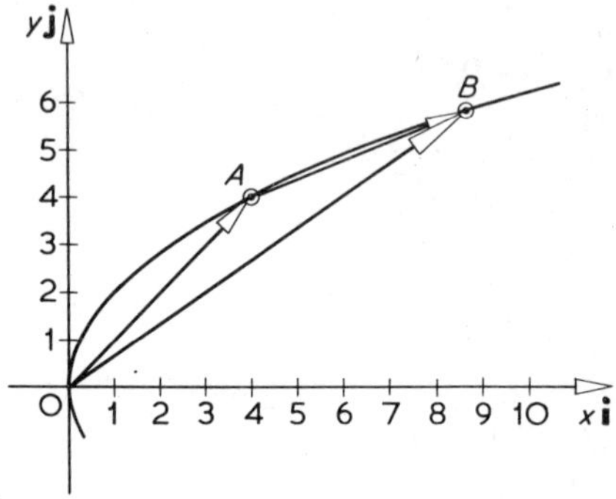

$\mathbf{AB} = 5\mathbf{i}+2\mathbf{j}$

4

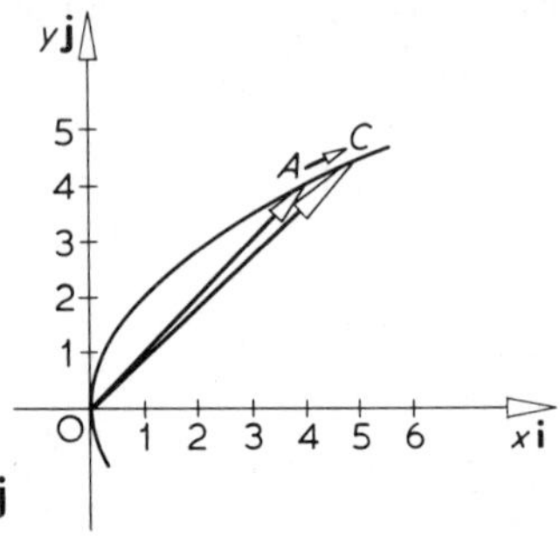

$$\frac{\mathbf{AC}}{0{\cdot}1} = 4{\cdot}1\mathbf{i}+2\mathbf{j}$$

whence the vector equation of the tangent at $p = 2$ is

$$\mathbf{r} = 4\mathbf{i}+4\mathbf{j}+t(4\mathbf{i}+2\mathbf{j})$$

5 (i) 10 seconds (ii) 490·5 metres (iii) 245·25 metres
(iv)

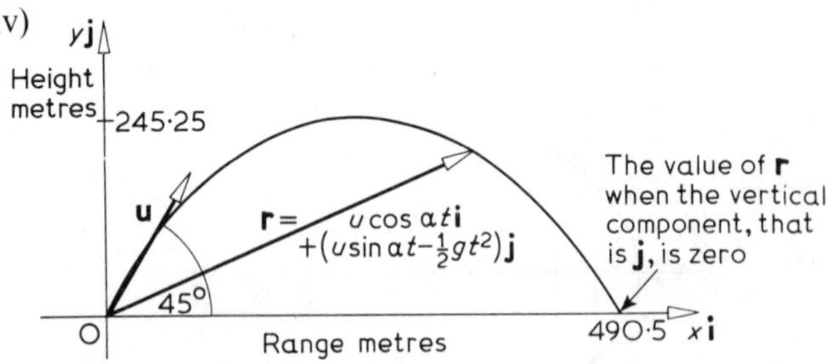

6 $\mathbf{r}_1 = 9\mathbf{i}+6\mathbf{j}$ $\mathbf{r}_2 = \mathbf{i}-2\mathbf{j}$

7 $\phi = \dfrac{p^2+3}{2p} = \dfrac{t^2+12}{4t}$

8 $p^2 = -2 \pm \sqrt{5}$ whence $p = \pm 0{\cdot}5692$ only
$\mathbf{r}_A = 0{\cdot}324\mathbf{i} + 1{\cdot}1384\mathbf{j}$, $\mathbf{r}_B = 0{\cdot}324\mathbf{i} - 1{\cdot}1384\mathbf{j}$ where A and B are the points of intersection. $AB = 2{\cdot}2768$

9 $a = -1$. When a has this value, $\mathbf{r} = \mathbf{i} + 2\mathbf{j}$ whence $r = \sqrt{5}$ units.

10 $8\sqrt{2}$ units.

Exercise 13

1 (*a*) $\mathbf{r} = 5\cos\phi\,\mathbf{i} + 3\sin\phi\,\mathbf{j}$
(*b*) $\mathbf{r} = 13\cos\phi\,\mathbf{i} + 5\sin\phi\,\mathbf{j}$

2 (i) (*a*) $(3\frac{3}{4}, 0), (-3\frac{3}{4}, 0)$ (*b*) $(12, 0), (-12, 0)$
(ii) (*a*) $x = \pm 6\frac{2}{3}$ (*b*) $x = \pm 14\frac{1}{12}$
(iii) (*a*) $3\frac{3}{5}$ (*b*) $3\frac{11}{13}$
(iv) (*a*) $\frac{3}{4}$ (*b*) $\frac{12}{13}$

3 (*a*) $\mathbf{r} = 5\cos\phi\,\mathbf{i} + 5\sin\phi\,\mathbf{j}$
(*b*) $\mathbf{r} = 13\cos\phi\,\mathbf{i} + 13\sin\phi\,\mathbf{j}$

4 (i) Ellipse, centre origin, major axis of 8 units on the **j** axis and minor axis of 6 units on the **i** axis.
(ii) Ellipse, centre (2, 1), major axis of 6 units parallel to the **i** axis and minor axis of 2 units parallel to the **j** axis.

5 $\phi = 2n\pi$, for integral values of n, including zero.

6 $\sqrt{6}$.

7 The particle follows an elliptical path, centred on the origin, major axis 4 units parallel to the **i** axis, minor axis 2 units parallel to the **j** axis. The time for one revolution is $\frac{1}{2}$ second. $t = (4n+1)/4$ for integral values of n, including zero.

8 $2\sqrt{3}$

9 $36° 52'$ $e = 4/5$

10 8 seconds

Exercise 14

1 These curves are two cycloids the first of which is formed by a point on the circumference of a circle radius 2 units which rolls along the x axis and the second is formed by a point on the rim of a circle of radius 1 unit.

2 (i) $\mathbf{r} = 3\pi\mathbf{i} + 6\mathbf{j}$
(ii) $\mathbf{r} = \frac{1}{2}(9\pi + 6)\mathbf{i} + 3\mathbf{j}$
(iii) $\mathbf{r} = \frac{3}{4}(7\pi + 2\sqrt{2})\mathbf{i} + \frac{3}{2}(2 - \sqrt{2})\mathbf{j}$

3 (i) Horizontal distance 3π units
Vertical distance 6 units

(ii) Horizontal distance $\frac{1}{2}(9\pi+6)$ units
Vertical distance 3 units
(iii) Horizontal distance $\frac{3}{4}(7\pi+2\sqrt{2})$ units
Vertical distance $\frac{3}{2}(2-\sqrt{2})$ units

4 (i) Horizontal distance $\frac{1}{2}(\pi-2)$ units
Vertical distance 1 unit
(ii) Horizontal distance π units
Vertical distance 2 units
(iii) Horizontal distance 2π units
Vertical distance zero.

5 $\mathbf{r} = \frac{1}{6}(2\pi-3)\mathbf{i}+\frac{\sqrt{3}}{2}\mathbf{j}$ Magnitude $\frac{1}{3}\sqrt{\pi^2-3\pi+9}$

Exercise 15

1

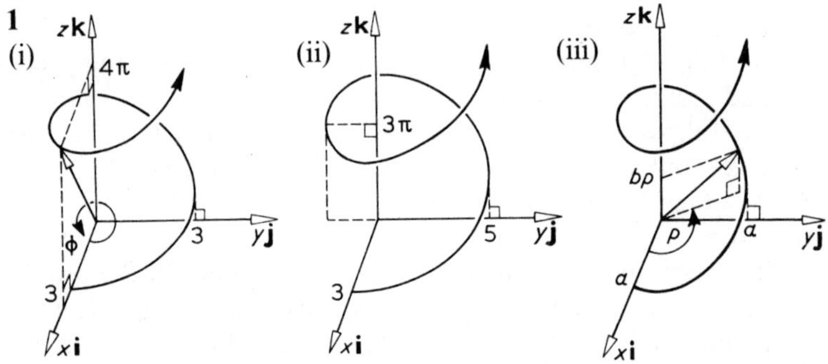

2 $\mathbf{r} = 4\cos 10\pi t\,\mathbf{i}+4\sin 10\pi t\,\mathbf{j}+\pi t\,\mathbf{k}$

3 $t = \pm a/5 = bp$

4 $2n\pi$ for integral values of n including zero.

5 $2\pi a$

6 For $\mathbf{r}_1$ and $\mathbf{r}_3$, $a = 1$ and $bp = 0$; for $\mathbf{r}_2$ and $\mathbf{r}_3$, $a = bp = 1$.

7 $\frac{1}{2}\sqrt{64+225\pi^2}$

8

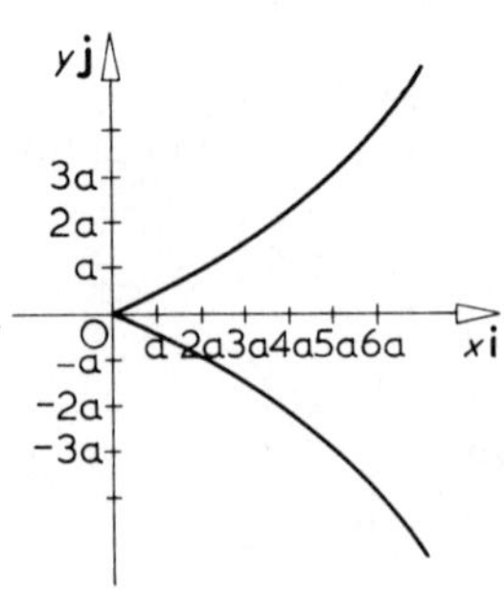

A semi-cubical parabola

9 An ellipse, centre origin, major axis of 6 units on the $\mathbf{i}$ axis and minor axis of 4 units on the $\mathbf{j}$ axis.

10 $\mathbf{r} = 4\cos\theta\,\mathbf{i} + 4\sin\theta\,\mathbf{j} + \pi\mathbf{k}$

Distance π units

$\mathbf{r} = 4\mathbf{i}(1-t) + 4t\mathbf{j} + \frac{9}{4}\pi t\mathbf{k}$

Exercise 16

1 (i) $\mathbf{v} = \cos t\,\mathbf{i} - \sin t\,\mathbf{j} + 2\cos 2t\,\mathbf{k}$

$\mathbf{a} = -\sin t\,\mathbf{i} - \cos t\,\mathbf{j} - 4\sin 2t\,\mathbf{k}$

(ii) $\mathbf{v} = 3\mathbf{i} + 2t\,\mathbf{j} - 9t^2\mathbf{k}$ $\mathbf{a} = 2\mathbf{j} - 18t\,\mathbf{k}$

(iii) $\mathbf{v} = (\cos t + \sin t)\mathbf{i} + (3\cos t - 2\sin t)\mathbf{j} + 2\sin 2t\,\mathbf{k}$

$\mathbf{a} = (\cos t - \sin t)\mathbf{i} - (3\sin t + 2\cos t)\mathbf{j} + 4\cos 2t\,\mathbf{k}$

(iv) $\mathbf{v} = 2e^{2t}\mathbf{i} - e^t\mathbf{j} + 6e^{3t}\mathbf{k}$ $\mathbf{a} = 4e^{2t}\mathbf{i} - e^t\mathbf{j} + 18e^{3t}\mathbf{k}$

(v) $\mathbf{v} = \mathbf{i} - (2t + 3t^2)\mathbf{j} - e^{-t}\mathbf{k}$ $\mathbf{a} = -(2+6t)\mathbf{j} + e^{-t}\mathbf{k}$

2 (i) $(t^2+3t)\mathbf{i} + (t-t^2)\mathbf{j}$

(ii) $(3-2t)\mathbf{i} - (2t+1)\mathbf{j}$

(iii) $6t$

(iv) $-(4t^3+6t)\mathbf{k}$

(v) $(4t^3+6t)\mathbf{k}$

3 (i) $\frac{1}{2}t^2\mathbf{i} + t^2\mathbf{j} + t\mathbf{k} + \text{const.}$

(ii) $-\cos t\,\mathbf{i} + \sin t\,\mathbf{j} + \text{const.}$

(iii) $\frac{1}{4}t^4 + \frac{1}{3}t^3 + \text{const.}$

4 (i) $\left(\frac{dx}{ds}\mathbf{i} + \frac{dy}{ds}\mathbf{j} + \frac{dz}{ds}\mathbf{k}\right) \Big/ \frac{dr}{ds}$

(ii) $\mathbf{R} = \mathbf{r} + u\hat{\mathbf{t}}$, for a parameter u.

5 (i) $k\hat{\mathbf{n}} = d\mathbf{t}/ds$ since $d\mathbf{t}/ds$ is perpendicular to $\mathbf{t}$; $\mathbf{t}$ is of constant length.

(ii) $(\mathbf{R}-\mathbf{r}).\mathbf{t} = 0$

6 $\mathbf{T} = \frac{1}{2}\{\mathbf{i} + \sqrt{3}\mathbf{j} + u(-\sqrt{3}\mathbf{i} + \mathbf{j})\}$

$\mathbf{N} = \frac{1}{2}\{\mathbf{i} + \sqrt{3}\mathbf{j} + u(\mathbf{i} + \sqrt{3}\mathbf{j})\}$

7 $\mathbf{T} = 5\mathbf{i} + (u-4)\mathbf{j}$

$\mathbf{N} = (5+u)\mathbf{i} - 4\mathbf{j}$

8 (i) $y = 0$

(ii) $x = y$

9 $\mathbf{T} = ap_1^2\mathbf{i} + 2ap_1\mathbf{j} + u(p_1\mathbf{i} + \mathbf{j})$

$\mathbf{N} = ap_1^2\mathbf{i} + 2ap_1\mathbf{j} + u\mathbf{i}$

10 (i) $(3\sqrt{2}/2)$ seconds

(ii) 22·0725 metres

(iii) $14{\cdot}715\sqrt{2}(\mathbf{i}-\mathbf{j})$

(iv) $14{\cdot}715\sqrt{2}\mathbf{i}$

11 $\mathbf{v}_{t=0} = 40\pi\mathbf{j}+\pi\mathbf{k}$ $\mathbf{a}_{t=0} = -400\pi^2\mathbf{i}$
$\mathbf{v}_{t=1} = 40\pi\mathbf{j}+\pi\mathbf{k}$ $\mathbf{a}_{t=1} = -400\pi^2\mathbf{i}$

12 $\mathbf{r} = 21\mathbf{i}+10\mathbf{j}+3\mathbf{k}$
$\dot{\mathbf{r}} = 36\mathbf{i}+11\mathbf{j}+\mathbf{k}$
$\ddot{\mathbf{r}} = 42\mathbf{i}+6\mathbf{j}$

13 $\mathbf{v}$: 21 $\mathbf{a}$: 26

14 $\dot{\mathbf{r}} = (\sin t+16)\mathbf{i}+(29-\cos t)\mathbf{j}+(16-gt)\mathbf{k}$
$\mathbf{r} = (1+16t-\cos t)\mathbf{i}+(29t-\sin t)\mathbf{j}+(16t-\tfrac{1}{2}gt^2)\mathbf{k}$

15 $\dot{\mathbf{r}} = a\cos\theta\,\dot{\theta}\,\hat{\mathbf{r}}+a\sin\theta\,\dot{\theta}\,\hat{\mathbf{q}}$
$\ddot{\mathbf{r}} = a(\cos\theta\,\ddot{\theta}-2\sin\theta\,\dot{\theta}^2)\hat{\mathbf{r}}+a(\sin\theta\,\ddot{\theta}+2\cos\theta\,\dot{\theta}^2)\hat{\mathbf{q}}$

16 (i) $v_r = \tfrac{1}{2}a$ $v_t = \tfrac{1}{4}at$
$a_r = \tfrac{1}{8}at$ $a_t = \tfrac{1}{2}a$

(ii) $v_r = -\dfrac{2a}{t^2}$ $v_t = \dfrac{a}{t}$

$a_r = \dfrac{4a}{t^3}-\dfrac{a}{2t}$ $a_t = -\dfrac{2a}{t^2}$

(iii) $v_r = 2at$ $v_t = at^2$
$a_r = 2a-at^2$ $a_t = 4at$

(iv) $v_r = -a\sin t$ $v_t = -2a\cos t$
$a_r = a\cos t$ $a_t = -2a\sin t$

17 $v_r = a\omega r$, for constant angular velocity ω, $v_t = r\omega$
$a_r = \omega^2 r(a^2-1)$ $a_t = 2a\omega^2 r$

18 $\mathbf{a} = \sqrt{4+r^2}\,\hat{\mathbf{t}}+k(r^2+4t^2)\hat{\mathbf{n}}$

19 $\ddot{\mathbf{r}} = -\omega^2\mathbf{r}$

20 $\dfrac{1}{\mu\theta} = \dfrac{1}{2\lambda r^2}+\text{const.}$

Radial acceleration $2\lambda^2 r^3-\mu^2\theta^4/r$

Transverse acceleration $\mu\theta^2\left(\lambda r+\dfrac{2\mu\theta}{r}\right)$

Revision exercises

I

1 (i) [7, −2, −3] (ii) [4, −2, −5] (iii) [3, 1, 0]
(iv) [1, −3, 0] (v) [3, 4, 4]

2 (i) $\dfrac{x-1}{1} = \dfrac{y}{-1} = \dfrac{z-3}{2}$

(ii) $\dfrac{x+1}{3} = \dfrac{y-2}{2} = \dfrac{z-1}{-2}$

(iii) $\dfrac{x-3}{2} = \dfrac{y-4}{-1} = \dfrac{z-5}{5}$

3 Since the components of $\mathbf{F}_1$ and $\mathbf{F}_2$ can never be equal, the two forces are never equal.

$a = 2 \qquad \mathbf{r} = -3\mathbf{i}-8\mathbf{j}-10\mathbf{k}+t(4\mathbf{i}+7\mathbf{j}+8\mathbf{k})$

4 $12\sqrt{2}\mathbf{i}+9\sqrt{2}\mathbf{j}$

5 Centre (1, 5, 6) Radius 5 metres Inclination 53° 8′

$\mathbf{v} = 3(\mathbf{i}+2\mathbf{j}+2\mathbf{k})$

II

1 $\mathbf{AB} = \mathbf{b}-\mathbf{a}$ and $\mathbf{AC} = 2(\mathbf{a}-\mathbf{b})$

Hence, $\mathbf{AC} = -2\mathbf{AB}$ and the result follows.

2 $\mathbf{r} = \mathbf{a}+p(\mathbf{b}-\mathbf{a})+q(\mathbf{c}-\mathbf{a})$ where p and q are real numbers.

3 4 J

4 $2\sqrt{3}\mathbf{i}+\mathbf{j}$

5 Charlie in $2\sqrt{1{\cdot}3}$ seconds Fred: 4 seconds $\mathbf{r} = 7\mathbf{i}+\mathbf{j}$

III

1 Shown by $k[1, -3, -4]$

2 (i) $x-z = 2$

(ii) $x+4y-5z = 6$

3 $\frac{1}{3}$

4 $\mathbf{r} = -\mathbf{i}+\frac{1}{4}\mathbf{k}$

Equation of circle $\mathbf{r} = \cos\theta\,\mathbf{i}+\sin\theta\,\mathbf{j}+\frac{1}{4}\mathbf{k}$

Successive distances are measured in a straight line parallel to the z axis.

5 Relative displacement

$$\mathbf{s}_1-\mathbf{s}_2 = t(\mathbf{a}-\mathbf{c})\mathbf{i}+t(\mathbf{b}-\mathbf{d})\mathbf{j} = t(\mathbf{v}_1-\mathbf{v}_2)$$

At 45°, $\dfrac{\mathbf{b}-\mathbf{d}}{\mathbf{a}-\mathbf{c}} = 1$, whence the result.

IV

1 $t[2, -1, 3]$

2 $2x-3y+2z-9 = 0$ Distance $8\sqrt{17}/17$

3 $\mathbf{v}_1 = 10\sqrt{3}\mathbf{i}+2\frac{1}{2}\mathbf{j}$ and $\mathbf{v}_2 = 7\frac{1}{2}\mathbf{j}$

4 Pitch $= \dfrac{\text{height through one revolution}}{\text{circumference}} = \dfrac{3}{\pi}$

$\mathbf{r}_1 = \mathbf{j}+\sqrt{3}\mathbf{k}+2\mathbf{i} \qquad \mathbf{r}_2 = -\mathbf{j}+\sqrt{3}\mathbf{k}+4\mathbf{i}$

5 Crossing point $100\mathbf{i}+100\mathbf{j}+30\mathbf{k}$ Time 10 seconds

V

1 Shown by $k[\mathbf{a}-3\mathbf{b}]$

2 (i)

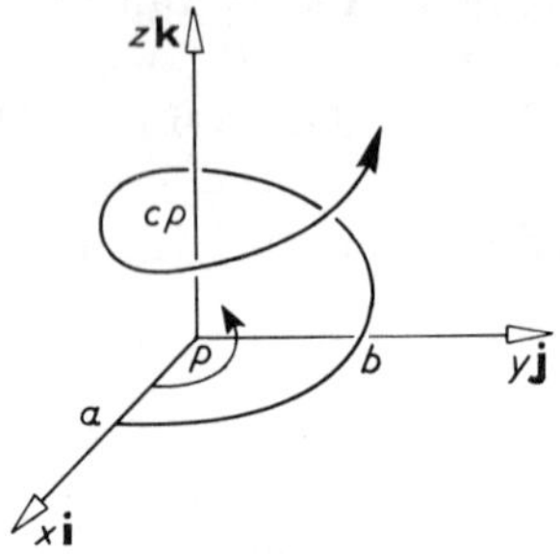

An elliptic helix

(ii) A circle radius a units, centre b units on the z axis and in a plane parallel to the plane $z = 0$.

3 $-\mathbf{i}-2\sqrt{3}\mathbf{j}$

4 $b = 3$

$\mathbf{r}_{z=2} = 10\cos\phi\,\mathbf{i}+6\sin\phi\,\mathbf{j}+2\mathbf{k}$

$\mathbf{r}_{z=n} = 5n\cos\phi\,\mathbf{i}+3n\sin\phi\,\mathbf{j}+n\mathbf{k}$

5 $\mathbf{F} = [3, 2, 4]$ 30 J

VI

1 $z = 0$

2 $\mathbf{v}_1 = 10\sqrt{3}\mathbf{i}+\mathbf{j}$ $\mathbf{v}_2 = 6\mathbf{j}$

3 Let $\mathbf{v}_A = gt\,\mathbf{j}+\mathbf{a}$ and $\mathbf{v}_B = gt\,\mathbf{j}+\mathbf{b}$ whence the relative velocity is $\mathbf{v}_B-\mathbf{v}_A = \mathbf{b}-\mathbf{a}$ = a constant vector.

4 Locus: $(2+3t+\frac{5}{8}t\cos\theta)\mathbf{i}+(-8+4t+\frac{5}{8}t\sin\theta)\mathbf{j}$

Time: $1\frac{3}{5}$ hours Position vector: $7\frac{2}{5}\mathbf{i}-\frac{4}{5}\mathbf{j}$

5 $\mathbf{r} = 4\cos 10\pi t\,\mathbf{i}+4\sin 10\pi t\,\mathbf{j}+\pi t\,\mathbf{k}$. Rate of increase is given by $2\pi\mathbf{k}$.

VII

1 $\mathbf{a} = k\mathbf{b}$, that is $[a_1, a_2, a_3] = k[b_1, b_2, b_3]$ which means that $a_1 = kb_1$, $a_2 = kb_2$ and $a_3 = kb_3$ simultaneously.

These points are not collinear since the $\mathbf{k}$ components are not in the same ratio as the $\mathbf{i}$ and $\mathbf{j}$ components.

2 $\dot{\mathbf{r}} = -a\sin t\,\mathbf{i}+a\cos t\,\mathbf{j}+b\mathbf{k}$

$\ddot{\mathbf{r}} = -a\cos t\,\mathbf{i}-a\sin t\,\mathbf{j}$

3 $\mathbf{OQ} = (3+2\cos\theta)\mathbf{i}+(4+2\sin\theta)\mathbf{j}+9\mathbf{k}$

Centre $(3, 4, 9)$ Radius 2.

4 $a = -5$, $\mathbf{b} = 1$, $p = -1$ and $q = 10$

Distance is 14 metres and so the work done is 196 J.

5 $\mathbf{v}_A = -\mathbf{i}-4\mathbf{j}$ $\mathbf{v}_B = 3\frac{2}{3}\mathbf{i}-3\mathbf{j}$

VIII

1 [2, 1, 1]
2 $\boldsymbol{\omega} = 2\mathbf{i}+3\mathbf{j}-\mathbf{k}$
3 $\lambda = 3$. Minimum length is zero since A coincides with B.
4 Displacement of projectile: $\mathbf{s} = at\,\mathbf{i}+(bt-\frac{1}{2}gt)\mathbf{j}$ whence $t = c/a$.
5 Time 36 minutes   Course due South

IX

1 $\overrightarrow{AB} = (4, -6, -4)$ $\overrightarrow{AC} = (2, -3, -2)$
Ratio: $-2{:}1$ and so an external division.
2 $a = \sqrt{2}$ $t = p = \pi/4$ or $5\pi/4$ Length: π units
3 $\boldsymbol{\omega} = 3\mathbf{i}+6\mathbf{j}-6\mathbf{k}$
4 Centre of both circles (1, 5, 6)   Radius of each 5 units.
Inclination of $\mathbf{r}_1$ 53° 8′, $\mathbf{r}_2$ 90°.
Line of intersection
$$45\sqrt{481}\mathbf{i}+60\sqrt{481}\mathbf{j}+80\sqrt{481}\mathbf{k}+t(\mathbf{i}+5\mathbf{j}+6\mathbf{k})$$
5 When $\tan\theta = \frac{1}{2}$, $u = 2v$ and the velocity vector of A is $v\mathbf{i}-v\mathbf{j}$.
When $\tan\theta = 7$, $u = \frac{1}{7}v$ and the velocity vector of A is $2\frac{6}{7}v\mathbf{i}-v\mathbf{j}$.

X

1 $\dfrac{d\hat{\mathbf{r}}}{dt} = \omega\hat{\mathbf{t}}$ $\dfrac{d\hat{\mathbf{t}}}{dt} = -\omega\hat{\mathbf{r}}$
2 $\dot{\mathbf{r}} = r\dot{\theta}(-\sin\phi\,\mathbf{i}+\cos\phi\,\mathbf{j})$
$\ddot{\mathbf{r}} = -r(\cos\phi\dot{\phi}^2+\sin\phi\ddot{\phi})\mathbf{i}+r(-\sin\phi\dot{\phi}^2+\cos\phi\ddot{\phi})\mathbf{j}$
3 $p = 2a(\mathbf{b}-\mathbf{a})/g$ $\tan\alpha = 3$
4 $\mathbf{s}_A = 9\mathbf{i}+7\mathbf{j}$ $\mathbf{s}_B = 3\mathbf{i}-\mathbf{j}$ Distance 10 nautical miles
5 Centre (2, 3, 4)   Radius 5 units
Maximum distance 7 units   Minimum distance 1 unit
Inclination 36° 52′

XI

1 $\dfrac{d\hat{\mathbf{u}}}{dt}$ is of magnitude ω in a direction at right angles to $\hat{\mathbf{u}}$ in the plane. $\omega\hat{\mathbf{t}}$
2 $\dfrac{d\mathbf{r}}{dt} = \dot{r}\hat{\mathbf{r}}+r\omega\hat{\mathbf{s}}$
Acceleration $(\ddot{r}-r\omega^2)\hat{\mathbf{r}}+(r\dot{\omega}+2\dot{r}\omega)\hat{\mathbf{s}}$
3 $\dot{\mathbf{r}} = \dot{r}\hat{\mathbf{r}}+r\omega\hat{\mathbf{t}}$ where $\hat{\mathbf{t}}$ is the unit vector perpendicular to $\hat{\mathbf{r}}$ and ω is the angular speed.
$\ddot{\mathbf{r}} = (\ddot{r}-r\omega^2)\hat{\mathbf{r}}+(r\dot{\omega}+2\dot{r}\omega)\hat{\mathbf{t}}$

4 $\frac{d}{dt}v\hat{\mathbf{t}} = \frac{dv}{dt}\hat{\mathbf{t}} + v\frac{d\hat{\mathbf{t}}}{dt} = \frac{dv}{dt}\hat{\mathbf{t}} + v\frac{d\hat{\mathbf{t}}}{ds}\frac{ds}{dt} = \frac{dv}{dt}\hat{\mathbf{t}} + v^2\frac{d\hat{\mathbf{t}}}{ds} = \frac{dv}{dt}\hat{\mathbf{t}} + kv^2\hat{\mathbf{n}}$

5 $\mathbf{a}_{t=0} = \frac{1}{\rho}\mathbf{N} + 2\mathbf{T}$ $\qquad \mathbf{a}_{t=3} = \frac{118}{\rho}\mathbf{N} + 2\sqrt{10}\mathbf{T}$

XII

1 $\hat{\mathbf{t}} = [-a\sin p, a\cos p, b]\frac{dp}{ds}$, where $\frac{dp}{ds} = \frac{1}{a^2+b^2}$

$k\hat{\mathbf{n}} = [-a\cos p, -a\sin p, 0]\frac{dp}{ds}\frac{1}{a^2+b^2}$

so that $\hat{\mathbf{n}} = [-\cos p, -\sin p, 0]$.

2 (i)

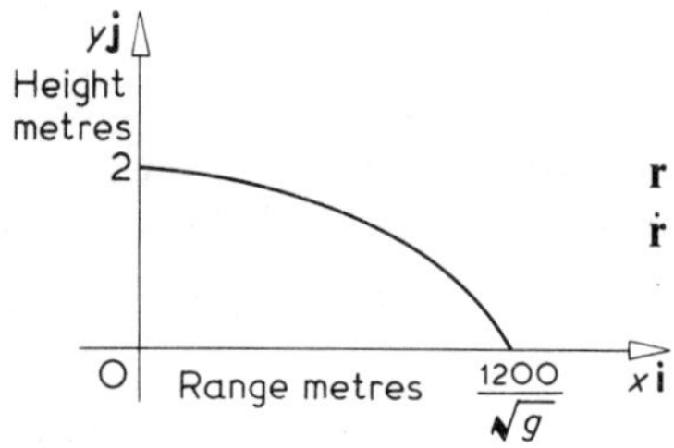

$\mathbf{r} = 600t\mathbf{i} - \frac{1}{2}gt^2\mathbf{j}$
$\dot{\mathbf{r}} = 600\mathbf{i} - gt\mathbf{j}$

(ii) $2/\sqrt{g}$ seconds
(iii) $1200/\sqrt{g}$ metres, horizontally.
(iv) $\dot{\mathbf{r}} = 600\mathbf{i} - 2\sqrt{g}\mathbf{j}$

3 (i) $\mathbf{v} = a\mathbf{i} + (b - gt\sin\theta)\mathbf{j}$

(ii) $\mathbf{r} = \frac{b}{2g\sin\theta}(2a\mathbf{i} + b\mathbf{j})$

Time $b/(g\sin\theta)$

4 $\mathbf{r}_1 = 2\mathbf{i} + \mathbf{j} + \sqrt{3}\mathbf{k}$
$\mathbf{r}_2 = 3\mathbf{i} + 2\mathbf{k}$
$\mathbf{r}_3 = 4\mathbf{i} - \mathbf{j} + \sqrt{3}\mathbf{k}$

5 $\mathbf{a} = 4b^2\mathbf{i} + 4b\mathbf{j}$

XIII

1 $\mathbf{T} = a(\cos p - u\sin p)\mathbf{i} + a(\sin p + \cos p)\mathbf{j} + b(p+u)\mathbf{k}$,
for a parameter u.
$\mathbf{N} = a\cos p(1-u)\mathbf{i} + a\sin p(1-u)\mathbf{j} + bp\,\mathbf{k}$

2 $\mathbf{v} = a\cos\theta\,\dot{\theta}\,\hat{\mathbf{r}} + a\sin\theta\,\dot{\theta}\,\hat{\mathbf{q}}$
$\mathbf{a} = a(\cos\theta\,\ddot{\theta} - 2\sin\theta\,\dot{\theta}^2)\hat{\mathbf{r}} + a(2\cos\theta\,\dot{\theta}^2 + \sin\theta\,\ddot{\theta})\hat{\mathbf{q}}$

3 $\ddot{r} = \frac{1}{3}r\omega^2 \qquad \dddot{r} = 3\dot{r}\omega^2$

4 $\mathbf{a} = \sqrt{4+r^2}\hat{\mathbf{t}}+k(r^2+4t^2)\hat{\mathbf{n}}$

5 **n** is in the **i**, **j** plane.

XIV

1 (*a*) Use either λ, μ method or $\mathbf{OP}-3\mathbf{OQ}+2\mathbf{OR} = \mathbf{0}$
that is $(\mathbf{OQ}+\mathbf{QP})-3\mathbf{OQ}+2(\mathbf{OQ}+\mathbf{QR}) = \mathbf{0}$

$$3\mathbf{OQ}+\mathbf{QP}-3\mathbf{OQ}+2\mathbf{QR} = \mathbf{0}$$

that is $\mathbf{QP} = -2\mathbf{QR}$ and the result follows.
(*b*) Since the components of **i** are both constant, the loci are both straight lines parallel to the $y\mathbf{j}$ axis and, therefore, parallel to each other.

$$\mathbf{OQ} = -2\mathbf{i}+1\tfrac{1}{3}\mathbf{j}$$

2 (i) $2\mathbf{AP} = \mathbf{AB}(1-k)+\mathbf{AC}(1+k)$
$2\mathbf{BQ} = \ldots \qquad 2\mathbf{CR} = \ldots$
Remember that $\mathbf{BA}+\mathbf{AC} = \mathbf{BC}$ in the subsequent addition and the result then follows.
(ii) $k\mathbf{PQ} = k(\mathbf{PA}+\mathbf{AQ}) = k(2\mathbf{BA}+1\tfrac{1}{2}\mathbf{AC})$
$= k(\mathbf{BA}+\tfrac{1}{2}\mathbf{AC}+\mathbf{BC}) = \ldots$
Compare coefficients and $p = 3k$, $q = 1\tfrac{1}{2}k$, $r = k$.

3 $OC = 13$ Angle $22^\circ 37'$

4 $(1+k)\mathbf{AX} = k\mathbf{AC}+\mathbf{AB}$
$(1+k)\mathbf{BY} = \ldots \qquad (1+k)\mathbf{ZE} = \ldots$
and after the subsequent addition the result follows in a similar way to the answer to question 2(i).

5 Since $\mathbf{AB}+\mathbf{BC}+\mathbf{CD}+\mathbf{DA} = \mathbf{0}$,

$$\mathbf{AB}+\mathbf{CD} = \mathbf{AD}+\mathbf{CB}$$

The resultant of **AB** and **CD** must act through the point of intersection of **AB** and **CD** and similarly for the resultant of **AD** and **CB** through the intersection of *AD* and *CB*. These resultants are equal and so the result follows.
$\mathbf{AB}+\mathbf{AD} = 2\mathbf{AE}$, $\mathbf{CB}+\mathbf{CD} = 2\mathbf{CE}$,
where *E* is the mid-point of *BD*.
$2\mathbf{AE}+2\mathbf{CE} = 4\mathbf{FE}$, where *F* is the mid-point of *AC* and this resultant acts through the mid-point of the line joining the intersections (*AD*, *BC*) and (*AB*, *CD*).

XV

1 $\mathbf{PA}+\mathbf{PB} = 2\mathbf{PN}$, where *N* is the mid-point of *AB*
$2\mathbf{PN}+\mathbf{PC} = 3\mathbf{PG}$, where $NG:GC = 1:2$

Therefore $\mathbf{PA}+\mathbf{PB}+\mathbf{PC} = \ldots$

Time $\frac{2\pi}{3}\sqrt{\frac{a}{3g}}$ Speed $\frac{1}{2}\sqrt{3ga}$

2 $p = -2$

$\mathbf{r} = 2(1+t)\mathbf{i}+(2+3t)\mathbf{j}+3\mathbf{k}$

3 The first and third ships collide at 2.20 p.m.

The other ship arrives at 2.30·77 p.m. from its position $13\frac{1}{3}\mathbf{i}+14\frac{1}{2}\mathbf{j}$ at 2.20 p.m.

4 When $n > 0$, $\mathbf{OP}_1 = a\cos 2\phi\,\mathbf{i}+a\sin 2\phi\,\mathbf{j}+a2\phi\,\mathbf{k}$

$\mathbf{OP}_2 = a\cos 4\phi\,\mathbf{i}+a\sin 4\phi\,\mathbf{j}+a4\phi\,\mathbf{k}$

Therefore

$$\mathbf{P}_1\mathbf{P}_2 = a(\cos 4\phi - \cos 2\phi)\mathbf{i}+a(\sin 4\phi-\sin 2\phi)\mathbf{j}+a2\phi\,\mathbf{k}$$

After simplification the result follows.

$$\mathbf{P}_1\mathbf{P}_3 = 2a\sqrt{\sin^2 2\phi+4\phi^2}$$

5 $\mathbf{a} = (f-c\,AP\sin\theta-c^2t^2AP\cos\theta)\mathbf{i}+(c\,AP\cos\theta-c^2t^2AP\sin\theta)\mathbf{j}$
$= \mathbf{0}$ for zero acceleration and the result follows. $c = \sqrt{2}/2$.

6 When $\tan\alpha = 1$, $u = V$ and $\mathbf{v}_A = a\mathbf{i}+b\mathbf{j} = V\mathbf{i}+V\mathbf{j}$

When $\tan\alpha = \frac{1}{3}$, $u = \frac{1}{3}V$ and $\mathbf{v}_A = 1\frac{2}{3}V\mathbf{i}+V\mathbf{j}$

XVI

1 $\mathbf{PO} = (1+\frac{2}{3}t)\mathbf{i}+\frac{2}{3}t\,\mathbf{j}+(2+\frac{1}{3}t)\mathbf{k}$ $\cos\theta = \dfrac{6-t}{\sqrt{6}\sqrt{3t^2-2t+6}}$

$\mathbf{PA} = \frac{2}{3}t\,\mathbf{i}+(\frac{2}{3}t-1)\mathbf{j}+(1+\frac{1}{3}t)\mathbf{k}$

2 $\mathbf{v} = 7\mathbf{i}+2\mathbf{j}$ Time 12.30 p.m. Position vector $4\frac{1}{2}\mathbf{i}+\mathbf{j}$

3 Find values for r when $\phi = 0, \pi/2, \pi \ldots$ to obtain major and minor axes and the result follows from symmetry about these axes or about the **i**, **j** plane where the projection of the curve is a circle. Distance is found by an inspection of the **j** components.

4 K.E. before impact $37\frac{1}{2}mu^2$. After impact $25mu^2$.

Along line of centres

$$(3+4)e = b-a \quad \text{and} \quad 6mu-4mu = 2amu+bm$$

5 Position vectors

first particle $nVt\cos\alpha\,\mathbf{i}+(nVt\sin\alpha-\frac{1}{2}gt^2)\mathbf{j}$

second particle $(h-Vt\cos\beta)\mathbf{i}+(k+Vt\sin\beta-\frac{1}{2}gt^2)\mathbf{j}$

Equate components, whence $\dfrac{k}{Vt}+\sin\beta = n\sin\alpha$ and the results follow.

$$V > \sqrt{\frac{gh}{2n\sin\alpha(n\cos\alpha+\cos\beta)}} \quad \text{or} \quad V > \sqrt{\frac{gk}{2n\sin\alpha(n\sin\alpha-\sin\beta)}}$$